Occupational Safety and Health Simplified for the Construction Industry

Occupational Safety and Health Simplified for the Construction Industry

Government Institutes
Research Group

Government Institutes

An imprint of
THE SCARECROW PRESS, INC.
Lanham, Maryland • Toronto • Plymouth, UK
2007

 **Government Institutes**

Published in the United States of America
by Government Institutes, an imprint of The Scarecrow Press, Inc.
A wholly owned subsidiary of
The Rowman & Littlefield Publishing Group, Inc.
4501 Forbes Boulevard, Suite 200
Lanham, Maryland 20706
http://www.govinstpress.com/

Estover Road
Plymouth PL6 7PY
United Kingdom

Contents

Preface

Occupational Safety and Health Simplified for the Construction Industry is written to help employers and employees handle the safety hazards they deal with on a construction site. This book is an effort to simplify in a single volume everything that most employers in the construction industry need to know about applicable Occupational Safety and Health Administration (OSHA) rules. It is a reference book, designed to be consulted frequently to help the reader understand and make sense of complicated OSHA regulations. However, it is important to note that it is a starting point only. Compliance with the law can only be achieved by consulting the actual Code of Federal Regulations (CFR). While this book is intended to help you make sense of the regulations, you must go to the regulations themselves in order to ensure compliance with the law.

Occupational Safety and Health Simplified for the Construction Industry

- covers how to prevent accidents from the most serious hazards in the construction industry, such as
  - electrical incidents,
  - falls,
  - struck-by, and
  - trenching and excavation;
- simplifies all of OSHA's Construction Standards Requirements;
- simplifies all of OSHA's Construction Training Requirements;
- simplifies how to communicate effectively with Spanish-speaking employees on your job site;
- includes a comprehensive easy-to-use checklist to conduct safety inspections on construction job sites; and
- includes a sample construction safety program to meet OSHA compliance requirements.

The book is written to help businesses that need the most current information in the construction industry and should be used as a reference book for quick answers to complicated questions.

It is the author's hope that this book will help employers prevent many of the construction injuries and fatalities that occur each year, while simplifying the OSHA compliance process. However, it is very important to remember that the only way you can comply with the law is to consult and comply with the rules in the Code of Federal Regulations. The applicable volumes of the CFR can be obtained from Government Institutes or the Government Printing Office.

CHAPTER 1

What Is the Occupational Safety and Health Act, and What Does It Require in the Construction Industry?

If you have one or more employees, you are an employer. If you are an employer, you are obligated to comply with the Occupational Safety and Health Act (OSH Act). There are very few exceptions. It doesn't just apply to manufacturing plants and construction sites, or only to hazardous jobs. It applies to all employers. Office work is covered. Retail stores are covered. If you employ anyone to do any kind of work, you must observe the OSH Act.

Occupational Safety and Health Administration (OSHA) requirements apply even if there are no accidents or injuries. The purpose of the OSH Act is to prevent the first injury. OSHA has penalized thousands of employers with perfect safety records. Other employers who have never been inspected by OSHA have been found liable in civil liability court cases because they had failed to observe an OSHA requirement. You must observe the OSHA requirements even though you are unaware of any hazardous conditions. It can be compared with the income tax rule that requires you to file a return even if you don't owe any taxes.

Employers are cited even when the OSHA violation is an employee's fault, such as failure to wear a hard hat, safety shoes, or goggles. OSHA places the onus on employers. The law provides that each employer is supposed to make sure that employees observe all safety requirements. There is no small business exception. While employers with ten or fewer employees don't have to keep injury/illness records and are exempted from some OSHA inspections, they must comply with all OSHA safety and health standards and requirements.

What Does the Law Require?

There are thousands of OSHA requirements. They include the following:

- Electricity rules
- Sanitation and air quality requirements
- Machine use, maintenance, and repair
- Posting notices and warnings

- Reporting accidents and illnesses
- Keeping detailed records
- Adopting written compliance program
- Employee training and qualifications

That's just a sample. There are several thousand others. OSHA violations occur when you do not observe all the requirements that apply. Simply running a safe operation is not enough. And ignorance of the law is no excuse. The penalties for noncompliance are severe. Million-dollar fines are not unusual. But that's not all. You could go to jail or be sued for millions even if you have never received an OSHA inspection.

Here's an Example

William Boss hired Joe Workman to do some painting. Boss didn't know that Workman had a slight asthmatic condition. Working with the paint severely aggravated Workman's condition. He had to be hospitalized and nearly died. He can never work again. Boss was sued for ten million dollars because he failed to heed the material safety data sheet (MSDS) warnings for the paint. They provided that no one should use the paint unless first given a pulmonary function test. The fact that Boss didn't know about the MSDS warnings or Workman's asthma was no defense. There is an OSHA requirement that an MSDS must be obtained for all products that contain hazardous chemicals and that each employer must provide training to his employees on the MSDS provisions. Boss hadn't done that.

Everyone can learn a lesson from this example. If you are required to provide protection, give a warning, provide employee training, or use particular compliance methods, and you have not complied because you didn't know about it, the consequences could be severe. If someone is hurt or becomes disabled as a result, you could face criminal charges or a multimillion-dollar liability suit, or both. And your defenses may be weak because ignorance of the law is no excuse. OSHA could also impose sanctions, even if no one was hurt.

The purpose of this book is to help you avoid problems of this kind.

What Are Our OSHA Compliance Obligations?

As an employer, your OSHA obligations fall into three general categories:

- In general, maintain a workplace free from recognized hazards that are likely to cause death or serious physical harm to your employees. That is a part of the OSH Act known as the "general duty clause." Its wording is rather ambiguous and indefinite. It is not used very often as the basis for citation, and it is not supposed to be. Congress directed OSHA to adopt specific standards to control workplace hazards.
- Observe all applicable occupational safety and health construction standards (OSHA standards) promulgated by the secretary of labor (29 CFR 1926). There are

thousands of OSHA standards. What they cover and who must comply with them is explained in this book. Most OSHA citations to date have alleged violations of OSHA standards. Understanding and observing them is, therefore, the most important of the three employer responsibilities listed here.

- For virtually every employer, keep records of employees on OSHA's Form 300. This is also known as the Log of Work-Related Injuries and Illnesses. It records injuries and illnesses; it reports work-related employee fatalities and multiple hospitalizations to OSHA and displays an OSHA-supplied poster that provides general information on the OSH Act.

Are OSHA State-Plan Standards Different than Federal OSHA Standards?

The OSHA obligations of employers located in the 23 state-plan states and territories are, for the most part, identical to those stated above. There are, however, some differences in some states.

The OSH Act requires that the OSHA standards in state-plan states be "at least as effective" as those adopted by OSHA itself. With few exceptions, the standards in those states are identical to OSHA's. They even use the same "CFR" designations.

Some state-plan states, however, have gone beyond OSHA in the adoption of regulations. They have adopted all OSHA standards and added some of their own. For example, there is no OSHA requirement that employers implement comprehensive workplace safety and health programs. But there are seven states that impose such a requirement: Alaska, California, Hawaii, Minnesota, Oregon, Texas, and Washington.

What Are the OSHA Construction Standards?

OSHA standards have the same status and effect as regulations adopted under other federal laws similar to the Internal Revenue Code, for example. You must comply with them, or you can be penalized with citations and fines.

Shortly after the OSH Act went into effect in 1971, the secretary of labor, under the authority delegated by Congress, adopted thousands of occupational safety and health standards. In subsequent years, additional standards have been added. Some of those standards have since been revised, while others have been revoked. The secretary's authority to adopt OSHA standards is a continuing one. Thus, new standards can be adopted in the future.

Job safety and health standards generally consist of rules for avoidance of hazards that have been proven by research and experience to be harmful to personal safety and health. The standards supposedly constitute an extensive compilation of wisdom. They sometimes apply to all employers, as do fire protection standards, for example. A great many standards, however, apply only to workers while engaged in specific types of work, such as driving a truck or handling compressed gases.

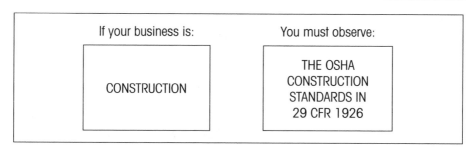

Figure 1.1. OSHA compliance requirements by industry

To demonstrate the form, the following is one of the many thousands of occupational safety and health standards: "Employees working in areas where there is a possible danger of head injury from impact, or from falling or flying objects, or from electrical shock and burns, shall be protected by protective helmets."

It is the obligation of all employers and employees to familiarize themselves with the standards that apply to them and to observe the standards at all times.

Once an OSHA standard has been adopted, it is published in the Code of Federal Regulations (CFR). The CFR is divided into 50 "titles" that cover all regulations adopted by all federal agencies. Each "title" is designated by a number, beginning with "1" and ending with "50." The OSHA standards are part of Title 29, the section of the CFR assigned to labor regulations.

Title 29 is further subdivided into various "parts" covering specific regulatory areas. All of OSHA's construction standards that apply to private employers are included in only one part: Part 1926—Occupational safety and health standards for construction.

Must I Have a Copy of the 1926 Construction Standards?

If federal OSHA administers your state, you must get a copy of the appropriate volume or volumes of the Code of Federal Regulations. Be sure to get a copy of the standards. There must be a copy physically present at every location where people work. If you do not have a copy of the standards, you will almost certainly be in violation of them. There are a number of OSHA standards, for example, that require the employer to make the text of the standard available for their employees. You are in violation of those standards if you do not have a copy available at each of your workplaces. See, for example, 29 CFR 1926.1020(g)(2).

It is not necessary that you read every standard, and even if you were so inclined, you almost certainly would not fully understand them. But you need to have them on hand. You need a copy of the OSHA construction standards in order to know that there are standards that cover particular aspects of your business. Knowing that is the first step. Understanding those standards comes later. But you cannot take the first step unless you have your own copy of the OSHA standards.

The OSHA standards are bound in paperback books. If you want all OSHA standards and regulations, you will have to get five separate volumes, but if you only want the construction standards, you'll only need one book.

How Can I Find Specific OSHA Construction Standards?

You don't have to read all the OSHA standards, but you will need to read particular standards at some point in time. Many OSHA standards are lengthy and complex. The difficulty of reading them is complicated by the use of small type print and the failure to separate each subsection by some readily identifiable method or to use a reader-friendly format.

All OSHA standards, however, are arranged in the same format or numbering system. Once that format is understood, locating particular subsections of a standard will be easier.

Publication of the Code of Federal Regulations (CFR) is the responsibility of the director of the Office of the Federal Register. They decide where in the CFR each federal agency's regulations will be published. Title 29 has been assigned to OSHA. Thus, each OSHA regulation is preceded by "29 CFR."

Individual parts of Title 29 are designated by a four-digit number (for example, 1926). That "part" designation is the first four numbers of all OSHA standards. A period follows the first four digits. Following that period, each standard is listed in numerical order beginning with number one. Thus 29 CFR 1926.1 refers to Title 29, Part 1926, and Section 1.

The number following the period is the designation given to a particular OSHA standard. In many cases, it will be followed by a name, for example, §1926.501 Duty to have fall protection. There will then follow various subsections of the standard. They are designated by letters and numbers, all of which are in parentheses. The first subsection is (a), for example, §1926.501(a). Other major subsections will be similarly designated in alphabetical order, for example, §1926.501(b), §1926.501(c), and so on. However, each of them often has its own subsections. And those subsections, in turn, have their subsections. The subsections are arranged and designated as follows:

- First group: alphabetically by small letters, for example, (a)
- Second group: numerically, for example, (1), (2), (3)
- Third group: numerically by roman numerals, for example, (i), (ii), (iii)
- Fourth group: alphabetically by capital letters, for example, (A), (B), (C)

For subsequent groups (if necessary), the order listed above is repeated, but the letter or number is in italics, for example, (*a*), (*1*), and so forth. An example to demonstrate follows.

Listed below are the steps to be taken to locate 29 CFR 1926.501(b)(7)(ii):

1. Get a copy of Title 29, CFR, Part 1926.
2. Locate §1926.501 Duty to have fall protection.
3. Begin with (a) until you find (b).
4. Find (1) under subsection (b). Then look for (7).
5. Once you have located (7), look for Roman numeral (i), then for (ii).

The format and sequence for all OSHA standards is the same. If these same steps are followed, you should be able to locate the particular subsection of the standard being sought.

CHAPTER 2

What Are the Most Serious Hazards in the Construction Industry?

Did you know?

One of every five workplace fatalities is a construction worker.

There are many hazards that can lead to injury in the construction industry. The hazards addressed in this chapter have been selected because statistics show they cause the most serious construction-related fatalities. Your company should focus on these areas to help ensure that potentially fatal accidents are prevented.

Nearly 6.5 million people work at approximately 252,000 construction sites across the nation on any given day. The fatal injury rate for the construction industry is higher than the national average for all industries. This chapter will help you identify and control the hazards that commonly cause the most serious construction injuries. They are grouped under the following headings:

- Falls
- Trenching and excavation
- Electric incidents
- Struck-by

Electrical Incidents

OVERVIEW OF OSHA'S ELECTRICAL STANDARD

Construction programs must address electrical incidents and the variety of ways electricity becomes a hazard. In general, the Occupational Safety and Health Administration (OSHA) requires that employees not work near any part of an electrical power circuit unless protected (29 CFR 1926.416[a][1]). Electricity has long been recognized as a serious workplace hazard, exposing employees to such dangers as electric shock, electrocution, fires, and explosions.

OSHA regulates electrical systems as used in construction under 29 CFR 1926, Subpart K, Electrical safety standards. Electrical safety is paramount, especially since electrical accidents are fairly common in construction.

Electrical regulations contained in OSHA's Construction Industry Standards (29 CFR 1926) address electrical safety requirements that are necessary for the practical safeguarding of employees involved in construction work. Subpart K is broken down into four sections for easier identification of the requirements:

- Installation safety requirements (1926.402–8): includes electric equipment and installations used to provide electric power and light on job sites.
- Safety-related work practices (1926.416 and 417): covers the hazards arising from the use of electricity at job sites, and arising from the accidental contact, direct or indirect, by employees with all energized lines, above or below ground, passing through or near the job site.
- Safety-related maintenance and environmental considerations (1926.431 and 432).
- Safety requirements for special equipment (1926.441).

Did you know?

Approximately 350 electrical-related fatalities occur each year.

The hazards discussed below are the most frequent causes of electrical injuries.

Burns

The most common shock-related injury is a burn. Burns suffered in electrical incidents may be one or more of the following three types:

- Electrical burns cause tissue damage and are the result of heat generated by the flow of electric current through the body. Electrical burns are one of the most serious injuries you can receive and should be given immediate attention.
- High temperatures near the body produced by an electric arc or explosion cause arc or flash burns. They should also be attended to promptly.
- Thermal contact burns occur when skin comes in contact with overheated electric equipment or when clothing is ignited in an electrical incident.

Contact with Power Lines

Overhead and buried power lines at your site are especially hazardous because they carry extremely high voltage. Fatal electrocution is the main risk, but burns and falls from elevations are also hazards. Using tools and equipment that can contact power lines increases the risk.

Overhead power lines that are uninsulated can carry tens of thousands of volts, making them extremely dangerous to employees who work in their vicinity.

Examples of equipment that can contact power lines follow:

- Aluminum paint rollers
- Backhoes
- Concrete pumpers
- Cranes
- Long-handled cement finishing floats
- Metal building materials
- Metal ladders
- Raised dump truck beds
- Scaffolds

Follow these procedures to avoid power line hazards:

- Look for overhead power lines and buried power line indicators. Post warning signs.
- Contact utilities for buried power line locations.

Case Studies: Deaths Due to Contact with Power Lines

The following incidents describe, in this order, situations in which a scaffold is too close to a power line; a crane boom is too close to a power line; and there is a lack of ground-fault protection.

Seven employees of a masonry company were erecting a brick wall from a tubular, welded-frame scaffold approximately 24 feet high. The scaffold had been constructed only 21 horizontal inches across from a 7,620-volt power line. A laborer carried a piece of wire reinforcement (ten feet long by eight inches wide) along the top section of the scaffold and contacted the power line with it. The laborer, who was wearing leather gloves, received an electric shock and dropped the wire reinforcement, which fell across the power line and simultaneously contacted the metal rail of the scaffold, energizing the entire scaffold. A 20-year-old bricklayer standing on the work platform in contact with the main scaffold was electrocuted.

A 56-year-old construction laborer was removing forms from a concrete wall poured several days earlier. As he removed the forms, he wrapped them with a length of cable called a choker, which was to be attached to a crane. The victim signaled the operator of the crane to extend the boom and lower the hoist cable. Both the operator and the victim failed to notice that the boom had contacted a 2,400-volt overhead power line. When the victim reached down to connect the choker to the hoist cable, he suddenly collapsed. Co-workers provided CPR but were unable to revive the victim. Only after a rescue squad arrived about four minutes later did anyone realize that the crane was in contact with a power line; all those present had assumed that the victim had suffered a heart attack.

Due to the dynamic, rugged nature of construction work, normal use of electrical equipment at your site causes wear and tear that results in insulation breaks, short circuits, and exposed wires. If there is no ground-fault protection, these can cause a ground-fault that sends current through the worker's body, resulting in electrical burns, explosions, fire, or death.

- Stay at least ten feet away from overhead power lines.
- Unless you know otherwise, assume that overhead lines are energized.
- De-energize ground lines when working near them. Use other protective measures include guarding or insulating the lines.
- Use nonconductive wood or fiberglass ladders when working near power lines.

How to Avoid Ground-Fault Protection Hazards

- Use ground-fault circuit interrupters (GFCIs) on all 120-volt, single-phase, 15- and 20-ampere receptacles, *or* have an assured equipment grounding conductor program (AEGCP).
- Follow manufacturers' recommended testing procedure to ensure the GFCI is working correctly.
- Use double-insulated tools and equipment, distinctively marked.
- Use tools and equipment according to the instructions included in their listing, labeling, or certification.
- Visually inspect all electrical equipment before use. Remove from service any equipment with frayed cords, missing ground prongs, cracked tool casings, and so forth. Apply a warning tag to any defective tool, and do not use it until the problem has been corrected.

Case Study:
Death Due to Lack of Ground-Fault Protection

A journeyman HVAC worker was installing metal ductwork using a double-insulated drill connected to a droplight cord. Power was supplied through two extension cords from a nearby residence. The individual's perspiration-soaked clothing and body contacted bare exposed conductors on one of the cords, causing an electrocution. No GFCIs were used. Additionally, the ground prongs were missing from the two cords.

PATH TO GROUND MISSING OR DISCONTINUOUS

If the power supply to the electrical equipment at your site is not grounded or the path has been broken, fault current may travel through a worker's body, causing electrical burns or death. Even when the power system is properly grounded, electrical equipment can instantly change from safe to hazardous because of extreme conditions and rough treatment.

How to Avoid Electrical Grounding Hazards

- Ground all power supply systems, electrical circuits, and electrical equipment.

- Frequently inspect electrical systems to ensure that the path to ground is continuous.
- Visually inspect all electrical equipment before use. Take any defective equipment out of service.
- Do not remove ground prongs from cord- and plug-connected equipment or extension cords.
- Use double-insulated tools.
- Ground all exposed metal parts of equipment.
- Ground metal parts of the following nonelectrical equipment, as specified by the OSHA standard 29 CFR 1926.404(f)(7)(v):
  - Frames and tracks of electrically operated cranes.
  - Frames of non–electrically driven elevator cars to which electric conductors are attached.
  - Hand-operated metal shifting ropes or cables of electric elevators.
  - Metal partitions, grill work, and similar metal enclosures around equipment of over 1kV between conductors.

Case Studies:
Deaths Due to Missing or Discontinuous Path to Ground

The following cases describe situations in which a ground wire is not attached and an adapter for a three-prong cord is not grounded to an outlet.

A fan connected to a 120-volt electrical system via extension cord provided ventilation for a worker performing a chipping operation from an aluminum stepladder. The insulation on the extension cord was cut through and exposed bare, energized conductors that made contact with the ladder. The ground wire was not attached on the male end of the cord's plug. When the energized conductor made contact with the ladder, the path to ground included the worker's body, resulting in death.

Two workers were using a 110-volt auger to install tie-down rods for a manufactured home. The auger has a one-quarter horsepower motor encased in a metal housing with two handles. One handle has a deadman's switch. Electricity to the auger was supplied by a series of 50-foot extension cords running to an adjacent property. Since the outlet at the adjacent property had no socket for a ground prong, the extension cords were plugged into the outlet using an adapter, but the ground wire of the adapter was not grounded. Two of the extension cords had no ground prongs, and some of them were repaired with electrical tape. The workers had removed their shirts and were sweating. One worker, holding the deadman's switch, received a shock from a ground-fault in the auger and was knocked back from the machine. The auger then fell across the other worker, the 24-year-old victim. The first worker knocked the auger off the victim but saw that the electric cord was wrapped around the victim's thigh. He yelled for his co-workers to disconnect the power, which they did. The workers administered CPR to the victim but to no avail.

EQUIPMENT NOT USED IN MANNER PRESCRIBED

If electrical equipment is used for a purpose for which it is not designed, you can no longer depend on safety features built in by the manufacturer. This may damage your equipment and cause injuries. Common examples of misused equipment follow:

- Using multireceptacle boxes designed to be mounted by fitting them with a power cord and placing them on the floor.
- Fabricating extension cords with Romex® wire.
- Using equipment outdoors that is labeled for use only in dry, indoor locations.
- Attaching ungrounded, two-prong adapter plugs to three-prong cords and tools.
- Using circuit breakers or fuses with the wrong rating for overcurrent protection, for example, using a 30-amp breaker in a system with 15- or 20-amp receptacles. Protection is lost because it will not trip when the system's load has been exceeded.
- Using modified cords or tools, for example, removing ground prongs, face plates, insulation, and so on.
- Using cords or tools with worn insulation or exposed wires.

Case Studies: Deaths Due to Misused Equipment

The following scenarios illustrate a damaged extension cord that leaves an arc welder ungrounded; handling a damaged extension cord when energized; and electrical equipment in poor condition.

A 29-year-old welder attempted to connect a portable arc welder to an electrical outlet using an extension cord. The power switch on the welder was already in the "on" position, and the female end of the extension cord, which was spring loaded, had apparently been dropped and broken. As a result, the ground prong of the welder plug did not insert into the ground terminal of the cord so that, as soon as a connection was made, the outside metal case of the welder became energized, electrocuting the victim. An examination revealed that the spring, cover plate, and part of the melamine casing were missing from the face of the female connector (the spring and some melamine fragments were found at the accident site). The victim was totally deaf in one ear and suffered diminished hearing in the other. He may have dropped the extension cord at the site and not heard the connector break.

A 19-year-old construction laborer was working with his foreman and another laborer to construct a waterfront bulkhead for a lakeside residence. Electricity for power tools was supplied from an exterior 120-volt, grounded AC receptacle located at the back of the residence. On the day of the incident, the victim plugged in a damaged extension cord and laid it out toward the bulkhead. There were no eyewitnesses to the accident, but evidence suggests that while the victim was handling the damaged and energized extension cord, he provided a "path to ground" and was electrocuted. The victim collapsed into the lake and sank four and a half feet to the bottom.

An 18-year-old worker at a construction site was electrocuted when he touched a light fixture while descending from a scaffold for his afternoon break. The source of the electricity was apparently a short in a receptacle, but examination revealed that the electrical equipment used by the contractor was in such poor condition that it was impossible to make a certain determination of the source of the short. Extension cords had poor splices, no grounds, and reversed polarity. One hand drill was not grounded, and the other had no safety plate. Out of several possible scenarios, the most likely was contact between the exposed wires of an extension cord and a screw that protruded from the receptacle, which had its face plate removed. The light fixture, which served as a ground, was known to be faulty for at least five months before the incident.

Improper Use of Extension and Flexible Cords

The normal wear and tear on extension and flexible cords at your site can loosen or expose wires, creating hazardous conditions. Cords that are not three-wire type, that are not designed for hard usage, or that have been modified increase your risk of contacting electrical current.

HOW TO AVOID EXTENSION CORD HAZARDS

- Use factory-assembled cord sets.
- Use only extension cords that are three-wire type.
- Use only extension cords that are marked with a designation code for hard or extra hard usage.
- Use only cords, connection devices, and fittings that are equipped with strain relief.
- Remove cords from receptacles by pulling on the plugs, not the cords.
- Continually audit cords on site. Any cords found not to be marked for hard or extra hard use, or which have been modified, must be taken out of service immediately.

Case Studies: Deaths Due to Improper Use of Extension and Flexible Cords

These two situations describe a flexible cord that is used instead of a three-wire, hard service cord, and a lack of strain relief.

A worker received a fatal shock when he was cutting drywall with a metal casing router. The router's three-wire power cord was spliced to a two-wire cord and plug set that was not rated for hard service. A fault occurred, and with no grounding and no GFCI protection, the worker was electrocuted.

A worker was operating a three-quarter electric chisel when an electrical fault occurred in the casing of the tool, causing him to be fatally electrocuted. An OSHA inspection revealed that the tool's original power cord had been replaced with a flat cord, which was not designated for hard service, and that strain relief was not provided at the point where the cord entered the tool. Additionally, the ground prong was missing and there was no GFCI protection.

Falls Protection Injuries and Fatalities

Did you know?

Falls from elevation account for one-third of all deaths in construction.

OVERVIEW OF OSHA'S FALL PROTECTION STANDARD

Each employee on a walking/working surface (horizontal and vertical surface) with an unprotected side or edge that is six feet or more above a lower level shall be protected from falls by the use of guardrail systems, safety net systems, or personal fall arrest systems. Each employee on walking/working surfaces shall be protected from falling through holes (including skylights) more than six feet above lower levels, by personal fall arrest systems, covers, or guardrail systems erected around such holes. Each employee on ramps, runways, and other walkways shall be protected from falling six feet or more to lower levels by guardrail systems.

In 2004, the Bureau of Labor Statistics (BLS) reported that 1,224 construction workers died on the job, with 36 percent of those fatalities resulting from falls. Events surrounding these types of accidents often involve a number of factors, including unstable working surfaces, misuse of fall protection equipment, and human error. Studies have shown that the use of guardrails, fall arrest systems, safety nets, covers, and travel restriction systems can prevent many deaths and injuries from falls.

Occupational fatalities caused by falls remain a serious public health problem. OSHA lists falls as one of the leading causes of traumatic occupational death, accounting for 8 percent of all occupational fatalities from trauma. Before you can begin a fall protection program, you must identify the potential fall hazards in your workplace. Any time a worker is at a height of six feet or more, the worker is at risk and needs to be protected.

It is important that safety and health programs contain provisions to protect workers from falls on the job. Unprotected sides, wall openings, and floor holes cause the most fall-related injuries.

Unprotected Sides, Wall Openings, and Floor Holes

Almost all sites have unprotected sides and edges, wall openings, or floor holes at some point during construction. If these sides and openings are not protected at your site, injuries from falls or falling objects may result, ranging from sprains and concussions to death.

How to Avoid Unprotected Openings Hazards

In general, it is better to use fall *prevention* systems, such as guardrails, than fall *protection* systems, such as safety nets or fall arrest devices, because they provide more positive safety means.

Whenever employees are exposed to a fall of six feet or more above a lower level, at least one of the following should be done:

- Use guardrail systems.
- Use safety net systems.

- Use personal fall arrest systems.
- Cover or guard floor holes as soon as they are created during new construction.
- For existing structures, survey the site before working, and continually audit as work continues. Guard or cover any openings or holes immediately.
- Construct all floor hole covers so they will effectively support two times the weight of employees, equipment, and materials that may be imposed on the cover at any one time.

Guardrail Systems

Where workers on a construction site are exposed to vertical drops of six feet or more, OSHA requires that employers provide fall protection in one of three ways *before* work begins:

- Place guardrails around the hazard area.
- Deploy safety nets.
- Provide personal fall arrest systems for each employee.

Many times the nature and location of the work will dictate the form that fall protection takes. If the employer chooses to use a guardrail system, he must comply with the following provisions:

- Top edge height of toprails, or equivalent guardrail system members, must be between 39 and 45 inches above the walking/working level, except when conditions warrant otherwise and all other criteria are met (e.g., when employees are using stilts, the top edge height of the toprail must be increased by an amount equal to the height of the stilts).
- Midrails, screens, mesh, intermediate vertical members, or equivalent intermediate structures must be installed between the top edge and the walking/working surface when there is no wall or other structure at least 21 inches high.
  - Midrails must be midway between the top edge of the guardrail system and the walking/working level.
  - Screens and mesh must extend from the toprail to the walking/working level and along the entire opening between rail supports.
  - Intermediate members (such as balusters) between posts must be no more than 19 inches apart.
  - Other structural members (such as additional midrails or architectural panels) must be installed so as to leave no openings wider than 19 inches.
- Guardrail systems must be capable of withstanding at least 200 pounds of force applied within two inches of the top edge, in any direction and at any point along the edge, and without causing the top edge of the guardrail to deflect downward to a height less than 39 inches above the walking/working level.
- Midrails, screens, mesh, and other intermediate members must be capable of withstanding at least 150 pounds of force applied in any direction at any point along the midrail or other member.

- Guardrail systems must not have rough or jagged surfaces that would cause punctures, lacerations, or snagged clothing.
- Toprails and midrails must not cause a projection hazard by overhanging the terminal posts.

Safety Net Systems

Where workers on a construction site are exposed to vertical drops of six feet or more, OSHA requires that employers provide fall protection by one of three ways *before* work begins:

- placing guardrails around the hazard area,
- installing safety nets, or
- providing personal fall arrest systems for each employee.

Many times the nature and location of the work will dictate the form that fall protection takes. If the employer chooses to use a safety net system, he must comply with the following provisions:

- Safety nets must be installed as close as practicable under the surface on which employees are working but in no case more than 30 feet below.
- When nets are used on bridges, the potential fall area must be unobstructed.
- Safety nets must extend outward from the outermost projection of the work surface as follows:

Table 2.1. Extension of safety nets

Vertical distance from working level to horizontal plane of net	Minimum required horizontal distance of outer edge of net from the edge of the working surface
Up to 5 feet	8 feet
5 to 10 feet	10 feet
More than 10 feet	13 feet

- Safety nets must be installed with sufficient clearance to prevent contact with the surface or structures under them when subjected to an impact force equal to the drop test described below.
- Safety nets and their installations must be capable of absorbing an impact force equal to the drop test described below.
- Safety nets and safety net installations must be drop tested at the job site
  - after initial installation and before being used,
  - whenever relocated,
  - after major repair, and
  - At six-month intervals if left in one place.
- The drop test consists of a 400-pound bag of sand 28–32 inches in diameter dropped into the net from the highest surface at which employees are exposed to fall hazards but not from less than 42 inches above that level.

- When the employer can demonstrate that it is unreasonable to perform the drop test described above, the employer or a designated competent person shall certify that the net and net installation have sufficient clearance and impact absorption by preparing a certification record prior to the net being used as a fall protection system. The certification must include
  - identification of the net and net installation;
  - the date that it was determined that the net and net installation were in compliance; and
  - the signature of the person making the determination and certification.
- The most recent certification record for each net and net installation must be available at the job site for inspection.
- Safety nets must be inspected for wear, damage, and other deterioration at least once a week, and after any occurrence that could affect the integrity of the system.
- Defective nets shall not be used, and defective components must be removed from service.
- Objects that have fallen into the safety net, such as scrap pieces, equipment, and tools, must be removed as soon as possible from the net and at least before the next work shift.
- Maximum mesh size must not exceed six inches by six inches. All mesh crossings must be secured to prevent enlargement of the mesh opening, which must be no longer than six inches, measured center to center.
- Each safety net, or section thereof, must have a border rope for webbing with a minimum breaking strength of 5,000 pounds.
- Connections between safety net panels must be as strong as integral net components and must not be spaced more than six inches apart.

Personal Fall Arrest Systems

A personal fall arrest system is one option of protection that OSHA requires for workers on construction sites who are exposed to vertical drops of six feet or more.

USING FALL ARREST SYSTEMS SAFELY

Ensure that personal fall arrest systems will, when stopping a fall,

- limit maximum arresting force to 1,800 pounds;
- be rigged such that an employee can neither free fall more than six feet nor contact any lower level;
- bring an employee to a complete stop and limit maximum deceleration distance to three and a half feet; and
- have sufficient strength to withstand twice the potential impact energy of a worker free-falling a distance of six feet, or the free-fall distance permitted by the system, whichever is less.

You should also remove systems and components from service immediately if they have been subjected to fall impact, until inspected by a competent person and deemed undamaged and suitable for use; promptly rescue employees in the event of a fall or ensure that they are able to rescue themselves; and inspect systems before each use for wear, damage, and other deterioration, and remove defective components from service. Also, do not attach fall arrest systems to guardrail systems or hoists, and rig fall arrest systems to allow movement of the worker only as far as the edge of the walking/working surface, when used at hoist areas.

IMPROPER SCAFFOLD CONSTRUCTION

Working with heavy equipment and building materials on the limited space of a scaffold is difficult. Without fall protection or safe access, it becomes hazardous. Falls from such improperly constructed scaffolds can result in injuries ranging from sprains to death.

PERSONAL FALL ARREST SYSTEMS

- Construct all scaffolds according to the manufacturer's instructions.
- Install guardrail systems along all open sides and ends of platforms.
- Use at least one of the following for scaffolds more than ten feet above a lower level: guardrail systems or personal fall arrest systems.

Do not use cross-bracing as a means of access.

Case Studies: Deaths Due to Improper Guardrails on Tubular Welded-Frame Scaffolds

The following reports on falls investigated by OSHA illustrate how seemingly innocent workplace activities can have deadly consequences.

An employee preparing masonry fascia for removal from a building fell from the third level of a tubular welded-frame scaffold. No guarding system was provided for the scaffold. Further, the platform was coated with ice, creating a slippery condition.

A contract employee was taking measurements from an unguarded scaffold inside a reactor vessel when he either lost his balance or stepped backwards and fell 14 1/2 feet, sustaining fatal injuries.

An employee was installing overhead boards from a scaffold platform consisting of two two-by-ten-inch boards with no guardrails. He lost his balance, fell seven and a half feet to the floor, and was fatally injured.

A laborer was working on the third level of a tubular welded-frame scaffold that was covered with ice and snow. Planking on the scaffold was inadequate; there was no guardrail and no access ladder for the various scaffold levels. The worker slipped and fell headfirst approximately 20 feet to the pavement below.

MISUSE OF PORTABLE LADDERS

You risk falling if portable ladders are not safely positioned each time they are used. While you are on a ladder, it may move and slip from its supports. You can also lose your balance while getting on or off an unsteady ladder. Falls from ladders can cause injuries ranging from sprains to death.

How to Avoid Ladder Hazards

- Position portable ladders so the side rails extend at least three feet above the landing.
- Secure side rails at the top to a rigid support, and use a grab device when three-foot extension is not possible.
- Make sure that the weight on the ladder will not cause it to slip off its support.
- Before each use inspect ladders for cracked or broken parts such as rungs, steps, side rails, feet, and locking components.
- Do not apply more weight on the ladder than it is designed to support.
- Use only ladders that comply with OSHA design standard 29 CFR 1926.1053(a)(1).

LADDER SAFETY

The OSHA Standard for portable ladders contains specific requirements designed to ensure worker safety:

Loads

Self-supporting (foldout) and non-self-supporting (leaning) portable ladders must be able to support at least four times the maximum intended load, except extra heavy-duty metal or plastic ladders, which must be able to sustain 3.3 times the maximum intended load.

Angle

Non-self-supporting ladders, which must lean against a wall or other support, are to be positioned at such an angle that the horizontal distance from the top support to the foot of the ladder is about one-quarter of the working length of the ladder. In the case of job-made wooden ladders, that angle should equal about one-eighth of the working length. This minimizes the strain of the load on ladder joints that might not be as strong as on commercially manufactured ladders.

Rungs

Ladder rungs, cleats, or steps must be parallel, level, and uniformly spaced when the ladder is in position for use. Rungs must be spaced between 10 and 14 inches apart.

For extension trestle ladders, the spacing must be 8–18 inches for the base, and 6–12 inches on the extension section. Rungs must be so shaped that an employee's foot cannot slide off and must be skid resistant.

Slipping

Ladders are to be kept free of oil, grease, wet paint, and other slipping hazards. Wood ladders must not be coated with any opaque covering, except identification or warning labels on one face only of a side rail.

Struck-By Injuries and Fatalities

Did you know?

One in four struck-by vehicle deaths involves construction workers.

The second highest cause of construction-related deaths is being struck by an object. Approximately 75 percent of struck-by fatalities involve heavy equipment such as trucks or cranes.

Struck-by injuries are a leading cause of both fatal and nonfatal injuries in the construction industry, yet much work is still needed to understand the nature of this diverse group of injuries. Using a data set of 3,390 injured construction workers treated at the George Washington University Emergency Department in Washington, D.C., injuries to workers who were struck by an object were studied.

To identify these injuries, researchers examined injuries in which the circumstance was coded as struck-by, struck against, caught, or cut, or where the injury involved a machine, an explosion, or a vehicle. To focus on those injuries where a worker was stuck by an object that was not under his control, researchers then removed injuries that were due to being caught between two objects, or where the worker was moving and hit himself on an object. There were 747 stuck-by injuries in this subset.

Researchers described in detail the nature and circumstances of these injuries. For example, researchers found that laborers were more likely to be involved in a "struck-by" injury (40 percent versus 29 percent for all workers combined). The injured worker was most frequently struck by a metal object (14 percent of all struck-by injuries), a pipe (8 percent), a board (7 percent), a beam (6 percent), a power tool (5 percent), and a scaffold (4 percent). A struck-by injury resulted more frequently in a contusion, a head injury, or a fracture than did other causes of injury.

If vehicle safety practices are not observed at your site, you risk being pinned between construction vehicles and walls, being struck by swinging backhoes, being crushed beneath overturned vehicles, or other similar accidents. If you work near public roadways, you risk being struck by trucks or cars.

HOW TO AVOID STRUCK-BY HAZARDS

- Wear seat belts that meet OSHA standards 29 CFR 1926.601(b)(9), except on equipment that is designed only for stand-up operation or that has no rollover protective structure.
- Check vehicles before each shift to ensure that all parts and accessories are in safe operating condition.
- Do not drive a vehicle in reverse gear with an obstructed rear view, unless it has an audible reverse alarm or another worker signals that it is safe.
- Drive vehicles or equipment only on roadways or grades that are safely constructed and maintained.
- Make sure that you and all other personnel are in the clear before using dumping or lifting devices.
- Lower or block bulldozer and scraper blades, end-loader buckets, dump bodies, and so on, when not in use, and leave all controls in neutral position.
- Set parking brakes when vehicles and equipment are parked, and chock the wheels if they are on an incline.
- All vehicles must have adequate braking systems and other safety devices.
- Haulage vehicles that are loaded by cranes, power shovels, loaders, and so forth, must have a cab shield or canopy that protects the driver from falling materials.
- Do not exceed a vehicle's rated load or lift capacity.
- Do not carry personnel unless there is a safe place to ride.
- Use traffic signs, barricades, or flaggers when construction takes place near public roadways.
- Workers must be highly visible in all levels of light. Warning clothing, such as red or orange vests, are required and, if worn for night work, must be of reflective material.

Case Studies: Deaths Due to Being Struck by a Vehicle

The following cases are reports of vehicle accidents investigated by OSHA and illustrate how seemingly innocent workplace activities can have deadly consequences.

An employee was operating a bulldozer at the top edge of a sloped drainage ditch. The bulldozer began to slide down the side of the snow and ice covered excavation, tipped over on its side, and pinned the operator under the roll bars. The driver was not wearing a seatbelt.

A contractor was operating a backhoe when an employee attempted to walk between the swinging superstructure of the backhoe and a concrete wall. As the employee approached from the operator's blind side, the superstructure hit the victim, crushing him against the wall. Employees had not been trained in safe work practices, and no barricades had been erected to prevent employee access to a hazardous area.

A safety "over travel" cable attached between the frame and the dump box of a dump truck caught on a protruding nut of an airbrake cylinder. This prevented the dump box from being fully raised. The driver, apparently assuming that releasing the cable would allow the dump box to continue upward, reached over the frame and disengaged the cable with his right hand. The dump box then dropped suddenly, crushing his head.

Falling and Flying Objects

You are at risk from *falling* objects when you are beneath cranes, scaffolds, and so forth, or where overhead work is being performed. There is a danger from *flying* objects when power tools, or activities like pushing, pulling, or prying, might cause objects to become airborne. Injuries can range from minor abrasions to concussions, blindness, or death.

How to Avoid Falling Hazards

POWER TOOLS AND MACHINES

- Use safety glasses, goggles, face shields, and so forth, where machines or tools may cause flying particles.
- Inspect tools, such as saws and lathes, to ensure that protective guards are in good condition.
- Make sure you are trained in the proper operation of powder-actuated tools.

CRANES AND HOISTS

- Avoid working underneath loads being moved.
- Barricade hazard areas and post warning signs.
- Inspect cranes and hoists to see that all components, such as wire rope, lifting hooks, chains, and so forth, are in good condition.
- Do not exceed lifting capacity of cranes and hoists.

Case Studies: Deaths Due to Falling or Flying Objects

The following cases are reports of accidents investigated by OSHA and illustrate how seemingly innocent workplace activities can have deadly consequences.

Two employees were using a wire rope to winch a wooden toolshed onto a flat bed trailer. The wire rope broke, snapped back, and struck one of the employees on the top of the head, killing him. The employee was not wearing a hard hat.

An employee was standing under a suspended scaffold that was hoisting a workman and three sections of ladder. Sections of the ladder became unlashed and fell 50 feet, striking the employee in the skull. The employee was not wearing any head protection and died from injuries received.

Workers were using a winch to pull a ten-foot section of a 600 lb. grain spout through a vent hole when the spout became wedged. Using pry bars, they attempted to free the spout, which was still under tension from the winch. When it popped free, the release of tension caused it to strike one of the workers in the head; the worker had no head protection.

A carpenter was attempting to anchor a plywood form in preparation for pouring a concrete wall using a powder-actuated tool. The nail passed through the hollow wall, traveled some 27 feet, and struck an apprentice in the head, killing him. The tool operator had never been trained in the proper use of the tool, and none of the employees in the area, including the victim, were wearing personal protective equipment.

Constructing Masonry Walls

Constructing concrete and masonry walls is especially dangerous because of the tremendous loads that need to be supported. There are risks of major accidents, and even death, when jacks or lifting equipment are used to position slabs and walls, or when shoring is required until structures can support themselves.

How to Avoid Masonry Wall Hazards

- Do not place construction loads on a concrete structure until a qualified person indicates that it can support the load.
- Adequately shore or brace structures until permanent supporting elements are in place or concrete has been tested to ensure sufficient strength.
- Only allow those who are essential to and actively engaged in construction or lifting operations to enter the work area.
- Take measures to prevent unrolled wire mesh from recoiling, such as securing each end or turning the roll over.
- Do not load lifting devices beyond their capacity.
- Use automatic holding devices to support forms in case a lifting mechanism fails.

Case Studies: Deaths Due to Constructing Masonry Walls

The following cases are reports of accidents investigated by OSHA and illustrate how seemingly innocent workplace activities can have deadly consequences.

Three concrete finishers were working in the basement of a home under construction, placing cement for the basement floor. A cement truck was parked two feet away from the west wall, unloading six yards of cement into the basement. The two-foot area around the foundation had been backfilled about an hour and a half before the cement finishers began their work. One of the employees directed the cement chute, starting at the northwest corner of the building. By the time he got to the southwest corner, the truck was empty. Suddenly, the west wall collapsed, crushing him to death. The other two employees were able to escape with only minor injuries.

In inclement weather, a 34-year-old worker was positioning vertical and horizontal rebar for a cap tie beam to be poured the next day. Potent, gusting winds caused a free-standing masonry block wall to collapse, fatally injuring the employee. Bracing and shoring could have prevented the collapse or lessened the impact.

The next victim was a member of a crew that was erecting tilt-up wall panels around the perimeter of the slab floor of a one-story warehouse. The last three wall slabs were being hoisted into place with two 12-foot nylon web slings in a basket hitch. While the second panel was suspended in preparation for being set, it tilted in the sling and slid slightly, cutting through one sling and partially through the other. The erection crew scattered as it dropped, but the victim stopped momentarily to look back as he fled the building. Just then, the upper edge of a previously set panel, which had been dislodged by the falling panel, fell on him. He was crushed and killed.

Trenching and Excavation Injuries and Fatalities

Did you know?

The fatality rate for excavation work is 112 percent higher than the rate for general construction work.

OVERVIEW OF OSHA'S EXCAVATION STANDARD

Cave-ins are perhaps the most feared trenching hazard. But other potentially fatal hazards exist, including asphyxiation due to lack of oxygen in a confined space, inhalation of toxic fumes, drowning, and so on. Electrocution or explosions can occur when workers contact underground utilities. OSHA requires that workers in trenches and excavations be protected and that safety and health programs address the variety of hazards they face.

The OSHA standards intend to protect workers in excavations and trenches. Basically, these standards require that walls and faces of all excavations in which workers are potentially exposed to danger from moving ground be guarded by a shoring system, safe sloping of the ground, or equivalent means of protection such as a trench shield or boxes. However, the standards are applicable only to trenches five feet or more in depth.

The standard applies to all open excavations made in the earth's surface, which includes trenches. OSHA defines a trench as a narrow excavation made below the surface of the ground in which the depth is greater than the width not exceeding 15 feet. An excavation is any man-made cut, cavity, trench, or depression in the earth's surface formed by earth removal. This can include excavations for anything from cellars to highways.

Excavating and trenching is recognized as one of the most hazardous construction operations. Employees in excavations must be protected from cave-ins except when the excavation is in stable rock, or is less than five feet deep, and an examination by a competent person provides no evidence that a cave-in should be expected.

Employees in excavations must also be protected from falling rock, soil, or material by use of an adequate system including scaling for loose rock or soil, installation of protective barricades, or other equivalent protection. Trench collapses cause dozens of fatalities and hundreds of injuries each year.

OSHA requires that workers in trenches and excavations be protected, and that safety and health programs address the variety of hazards they face. The following hazards cause the most trenching and excavation injuries:

NO PROTECTIVE SYSTEM

All excavations are hazardous because they are inherently unstable. If they are restricted spaces they present the additional risks of oxygen depletion, toxic fumes, and water accumulation. If you are not using protective systems or equipment while working in trenches or excavations at your site, you are in danger of suffocating, inhaling toxic materials, fire, drowning, or being crushed by a cave-in.

HOW TO AVOID TRENCHING AND EXCAVATION HAZARDS

Prejob planning is vital to accident-free trenching; safety cannot be improvised as work progresses. The following concerns must be addressed by a competent person:

- Evaluate soil conditions per 29 CFR 1926.652(P)(AppA) and select appropriate protective systems per 29 CFR 1926.652(P)(AppF).
- Construct protective systems in accordance with the standard requirements 29 CFR 1926.652.
- Preplan: contact utilities (gas, electric) to locate underground lines, plan for traffic control if necessary, and determine proximity to structures that could affect choice of protective system.
- Test for low oxygen, hazardous fumes, and toxic gases, especially when gasoline engine-driven equipment is running or the dirt has been contaminated by leaking lines or storage tanks. Ensure adequate ventilation or respiratory protection if necessary.
- Provide safe access into and out of the excavation.
- Provide appropriate protections if water accumulation is a problem.
- Inspect the site daily at the start of each shift, following a rainstorm, or after any other hazard-increasing event.
- Keep excavations open the minimum amount of time needed to complete operations.

FAILURE TO INSPECT TRENCH AND PROTECTIVE SYSTEMS

If trenches and excavations at your site are not inspected daily for evidence of possible cave-ins, hazardous atmospheres, failure of protective systems, or other unsafe conditions, you are in danger. Inspect excavations

- before construction begins,
- daily before each shift,
- as needed throughout the shift, and
- following rainstorms or other hazard-increasing events (such as a vehicle or other equipment approaching the edge of an excavation).

Inspections must be conducted by a competent person who

- has training in soil analysis,
- has training in the use of protective systems,
- is knowledgeable about the OSHA requirements, and
- has authority to immediately eliminate hazards.

UNSAFE ACCESS AND EGRESS

To avoid fall injuries during normal entry and exit of a trench or excavation at your job site, ladders, stairways, or ramps are required. In some circumstances, when conditions in a trench or excavation become hazardous, survival might even depend on how quickly you can climb out.

How to Avoid Access and Egress Hazards

- Provide stairways, ladders, ramps, or other safe means of egress in all trenches that are four feet deep or more.
- Position means of egress within 25 lateral feet of workers.
- Make sure that structural ramps that are used solely for access or egress from excavations are designed by a competent person.
- When two or more components form a ramp or runway, connect them to prevent displacement and to be of uniform thickness.
- Attach cleats or other means of connecting runway components in a way that would not cause tripping (e.g., to the bottom of the structure).
- Structural ramps used in place of steps must have a nonslip surface.
- Use earthen ramps as a means of egress only if a worker can walk them in an upright position and only if they have been evaluated by a competent person.

Case Studies: Deaths Due to Unsafe Access or Egress

The following cases are reports of trenching accidents investigated by OSHA and illustrate how seemingly innocent workplace activities can have deadly consequences.

Two employees were laying pipe in a trench 12-feet deep when one of the employees saw the bottom face of the trench move. He jumped out of the way along the length of the trench; the other employee was fatally injured as the wall caved in. The walls of the trench were not sloped, and no means of emergency egress were provided.

In a 15-foot-deep trench, which was not shored or sloped properly, two workers were laying sewer pipe. The only means of egress was by climbing the backfill. While exiting the trench, one of the workers was trapped by a small cave-in. The second employee tried to extricate him, but a second cave-in occurred, trapping the second employee at the waist. The second cave-in actually caused the death of the first employee; the second employee sustained a hip injury.

CHAPTER 3

How to Prevent Serious Injuries to Your Spanish-Speaking Employees in the Workforce

This chapter deals with employers who are hiring the growing number of Hispanic workers in the construction industry and demanding bilingual job safety and health training. This chapter will help provide employers with information on how to implement an effective safety and health program for their Hispanic workers.

The number of Hispanics living in the United States has increased by almost 60 percent in the last ten years, and projections are that, by the year 2007, Hispanics will account for 15 percent of the population, making them the country's largest minority. This growth has occurred not only in border states and eastern cities that are typically targets for Hispanic and Latino immigrants but also in the Midwest and Southeast where Spanish-speaking arrivals have gravitated toward the construction industry. In the Carolinas, for example, some 80 percent of construction workers are Hispanic.

Is the construction industry taking steps to accommodate the influx of Spanish and other non-English-speaking workers? Hispanics now comprise more than 20 percent of construction employees. The Hispanic proportion of the construction industry workforce has grown rapidly, and this will continue. This raises some of the most important safety issues facing the construction industry.

The Occupational Safety and Health Administration (OSHA) just recently fined a contractor $50,000 when a motorized compactor run by a non-English-speaking operator overturned, catching him under the rollover protection system and amputating his leg. Not only did the operator not speak English, but company supervisors did not speak Spanish either, and instructions for the compactor were in English only. OSHA found that because the employee wasn't properly trained in the limitations of his equipment, he entered an area that was too steep for the compactor to remain upright.

Success Story

One very successful story is the Dallas/Fort Worth Airport (DFWA) expansion project. The airport's safety training program (STP), a 40-hour training course, appears to be breaking down barriers of language, literacy, and culture and having a big impact on safety.

Such success couldn't come at a better time for Texas Hispanic construction workers. In 2003, 277 Hispanic workers lost their lives in construction-related incidents; 81 of these fatalities occurred in Texas, while 36 took place in California and 20 in Florida.

Such statistics prompted the Hispanic Contractors Association (HCAT), an association formed to recently declare a "state of emergency" for Hispanic construction workers in Texas.

The supervisor on the construction site is clear about what is seen as the solution to this problem. It's about training, communication, and culture. The main task for these workers is to train and provide information in the proper language before they need it.

An Admirable Safety Record

If Texas is facing tough challenges in the safety of its Hispanic workers, DFWA's Capital Development Program (CDP), as the airport expansion is called, may have one of the best construction training programs in the country. The program, which began in September 2002 and will entail 23 million man-hours of construction, has an admirable safety record. In addition to an injury rate far below the national average for a heavy construction site, DFWA is saving additional money on its project-controlled insurance program (PCIP). The average cost of a workers' compensation claim is more than 15 percent lower than the Texas average.

OSHA's area director for the Fort Worth office says, "It's a great program at DFWA." He thinks the program is unique because of the education they require everyone to have. Even on larger jobs they don't see that.

An OSHA inspection of the site one year ago turned up just 16 companies with violations of agency safety rules, out of 180 contractors on site at the time. That's pretty low. Usually, they find about 50 percent of contractors or subcontractors have a violation. Most of the violations had to do with unsafe acts, rather than unsafe conditions: employees who had unhooked fall protection equipment, or were under a suspended load.

The safety training program is not the only reason DFWA has compiled a good safety record. One of the biggest challenges in construction safety is inducing smaller subcontractors to take safety as seriously as larger companies. But by utilizing a "wrap-up" insurance program, the airport project enforces a single, universal safety program for all subcontractors at the worksite.

OSHA credits the universal safety and incentive programs as important contributing factors to DFWA's low injury rates. But what is most unique about DFWA's safety program is its mandatory 40-hour, bilingual safety training program. What lessons does it teach those who want to improve the safety of non-English-speaking construction workers?

The safety training program was developed by Best Institute Inc. of Garland, Texas. In conjunction with the two primary contractors at the airport project,

13,000 construction workers have taken the course and roughly half took it in Spanish.

Speak the Workers' Language

According to OSHA's interpretation of the Hazard communication standard, when employers have a training requirement, they must provide it in a language the worker can understand. Teaching in the appropriate language, however, is only the beginning. Successful training of Latino workers must be sensitive to differences in culture and education that distinguish Latinos from other workers and that even divide Latinos among themselves. Really reaching workers, affecting their behavior and attitudes, entails more than language fluency.

Part of the success of the program has been to recruit instructors who are from the ethnic groups being trained. Both instructors and curriculum developers are bilingual. The program also tries to use instructors who have worked in the construction industry.

Part of the training involves helping English-speaking and Spanish-speaking workers to understand basic construction terms. The 40-hour course doesn't try to make you fluent in the other language, but it does teach you to say "peligro" [danger] or "cuidado" [careful], if you see someone with his back to dangerous equipment. Knowing a few key words could save someone's life.

The classroom instruction is backed up with printed material workers can take on the job. They give workers these cards with Spanish to English on one side, and English to Spanish on the other.

Address Cultural Differences

The language barrier is the biggest challenge in employing Hispanic workers. It's a real plus if the Spanish-speaking person is bilingual. Anyone who is hiring laborers from the Hispanic community is probably hiring bilingual workers whenever they can, and these people are probably on a faster track than those without English.

It is very difficult to place workers in unionized jobs if Spanish is their only language. There are certain companies willing to work with workers who only speak Spanish, foundation companies, for example, until their English gets better, but for jobs like crane operators, where communication is important, they have to be bilingual.

Such observations aside, the fact remains that bilingual workers, whether laborers or supervisors, are hard to come by, a situation that makes it even more difficult to provide non-English-speaking employees the safety and operating information they need to protect themselves and their co-workers.

The need to reach Spanish-speaking workers is acute in a number of areas around the country. There are a lot of horror stories out there in states like Texas, New Mexico, and Arizona about people working under unsafe conditions and being put on equipment and told to learn on the job.

This makes the language barrier not the only problem. There is also the basic question of whether employers are providing training at all or adhering to safe work practices. In these circumstances, workers see things they know are not safe, but they're afraid to speak up because they think they'll lose their job.

In Florida, for example, where injuries and fatalities among Hispanic workers have made headlines and unionized employees are in the minority, many Spanish-speaking workers are employed by small companies where training and benefits, such as workers' comp and health insurance, are practically nonexistent.

These factors suggest that many employers haven't caught up with the fact that Hispanics are likely to be in the workplace for years to come. While OSHA doesn't explicitly require that employers train non-English-speaking workers in their native language, there is an awareness that workers for whom English isn't their native language might not get the same exposure to safety and health issues as do unionized or English-speaking workers.

Cultural differences between Hispanic and English-speaking workers can have huge safety implications. OSHA contends Hispanic workers are often very loyal and dedicated, and explains some differences between them and American workers. For example, if a hammer falls apart, you'll find the Latino has found a way to tape it back together, whereas an American worker will come to you and say, "You gave me this piece of crap and I can't do my job!"

Whatever the reason, the reluctance of many Latinos to challenge authority means they may agree to do unsafe jobs or not stop co-workers from risky behavior. This cultural aversion to saying no may well be one factor behind the high fatality rates for Hispanic workers.

Employees learn through this 40-hour course that they won't get fired for reporting unsafe acts or conditions. To encourage this kind of behavior, the DFWA set up a hotline so workers who call can do so confidentially.

Many Spanish workers on the project now feel comfortable about approaching others to remind them to wear safety glasses or hard hats, even those who are "higher up." But for this practice to be successful, cultural differences—even among Latinos—must be understood and respected.

In order to alter ingrained cultural and work practices, a 24-hour course is the absolute minimum, although a longer period is preferable. The first day, workers are usually still closed-minded, but by the third day, they start to see attitudinal change, especially because of nontraditional instructional methods. The 40-hour course used at DFWA is not cheap; it costs about $500 per student.

Verify Learning

Any experienced teacher knows you should never assume just because a lesson has been delivered that the information has been understood.

Well aware of the need to verify that a lesson has been learned, the training company figured out how to turn a problem into a solution. They couldn't use written tests due to the literacy problem, so they took a hands-on approach for adults.

Workers at the airport project learn by doing. The instructors demonstrate a skill such as the proper use of fall protection equipment. Then the workers duplicate the lesson in a special classroom until they get it right.

Follow-up

No training course, however effective, can provide permanent inoculation from occupational hazards. They now have a continuous quality improvement process that looks at how the training has affected employees on the line, to gauge the strengths and weaknesses.

Weekly safety meetings are also used to reinforce lessons and to address new job hazards as they emerge. In addition, workers and subcontractors are coached and educated to follow safety rules. They are disciplined, or even terminated, if they do not.

They have also developed a "pretask plan" means of communicating to their employees, sometimes in writing, before doing a job, so they understand the work better. This is always done in English and Spanish, which is something that is important to this project.

Wider Application

The expansion of DFWA is a big, publicly funded construction project. Can private companies use DFWA's ambitious training program, and is it applicable to smaller jobs? Because the airport has a high level of self-insurance it can save money directly through lower injuries and workers' compensation costs. Other large projects that can afford to self-insure are looking for ways to cut injuries and save money. Companies that can control losses help production and improve quality.

Companies that aren't large don't think they can do this. They might do a big job and expect losses and don't look at controlling accidents as a potential source of profit. Even one serious incident can result in higher workers' compensation for three years, but the cost of the problem, as well as the benefit of avoiding it, is delayed. The cost of an ambitious training program, however, is immediate.

How to Establish an Effective Training Program for Hispanic Workers

As discussed earlier, job training for limited-English Hispanic workers in many workplaces is often problematic. The trainer, usually a native English speaker without bilingual skills, grabs a native Spanish speaker with some English skills from the group of workers and uses this person as an interpreter.

The training begins with the trainer speaking through the interpreter. Sometime during the training session, the trainer asks the trainees if there are any questions. The trainees hardly ever have any questions. The training ends with the trainer asking the trainees if they understood everything. The trainees nod their heads in a yes motion, indicating they understood everything. But probably they did not.

Job safety training between people of different languages and cultures can be done effectively by following these steps:

1. Form an effective training team;
2. Communicate the company's belief that worker safety is important;
3. Initiate worker participation; and
4. Conduct safety training.

Step 1: Form an Effective Training Team

To be effective, the first step is to form an effective training team. Newly hired Spanish-speaking workers must learn many things quickly. They must learn how to do the job, they must learn how to act safely, and they must learn the workplace's unique culture. It is important for owners who hire Hispanic workers with limited English skills to know that because of this language barrier, these workers need more help than native English-speaking workers in learning how to do the job, how to act safely, and how to be consistent with the owner's beliefs and values.

Form an effective training team consisting of the following key people:

- Language leader—the person with the best bilingual skills.
- Social leader—the person the group recognizes as their leader.
- Technical leader—the person with the best skills and knowledge to get the job done.

The language leader can be a worker, a foreman, a supervisor, or a professional interpreter. If the job is complex and requires multiple safety measures, it may be very important to obtain the services of a professional interpreter. Without the language leader as a member of the training team, communication will likely be limited to hand signs. If you are planning on using one of the workers as the interpreter, identifying the language leader is easy. Simply ask the group of workers, "Who speaks the best English?" In most cases, someone in the group will volunteer to speak English.

The social leader is usually one of the workers. This person might be the individual who helped the workers get to the United States from Latin America. It's usually someone in high standing in the group's home village or, simply, the person recognized as the one who makes the important group decisions or who influences others. Without the social leader as a member of the training team, the group might not trust what is being communicated. If the social leader is excluded from the training team and he or she and the language leader don't trust each other, the training session could be undermined by the social leader with just a few glances to co-workers. Identifying the social leader is also easy. If the group has been together for a period of time, the social leader is probably already known to you. If not, ask the group,

"Who is your leader?" In most cases, someone in the group will identify who is the leader of the group.

The technical leader can be the owner, the foreman, the supervisor, or the person with the most experience. The technical leader must have two important qualities: he or she must know how to do the job well, and he or she must be able to do it safely. Without the technical leader as a member of the training team, the best and safest way to do the job will not be communicated. It is harder to relearn to do something right than to learn to do it right the first time. Identifying the technical leader is also easy. Who is the person that can do the job the best, the most efficiently, and most safely?

Congratulations, you have identified the three key people in your training team!

Step 2: Communicate the Company's Belief in the Importance of Worker Safety

For the second step, the company's owner must recognize that worker safety is as important as production and effectively communicate this belief. This is accomplished through establishing an effective accountability system. Managers, supervisors, and foremen actions must be measured by meeting both safety and production goals.

When someone doesn't speak the language, much of the learning takes place visually. For this reason, a company hiring limited-English-speaking workers must ensure that safe work procedures are already in place.

For example, a newly hired Spanish-speaking worker sees a co-worker walk through the construction site without wearing his hard hat. The co-worker stops to chat to a foreman who is also not wearing his hard hat. The message is clear: you don't have to wear a hard hat on this job site!

Step 3: Initiate Worker Participation

The following are three ideas to encourage worker participation:

1. Have a conversation about cultural attitudes toward safety;
2. Establish a safety award program; or
3. Implement a hazard identification contest.

HAVE A CONVERSATION ABOUT CULTURAL ATTITUDES TOWARD SAFETY

Asking someone's opinion makes them feel valued. Here are safety questions to ask that will make workers the experts and will teach you about them:

- What rights do workers have in your country regarding a safe work environment?

- During the time you have been working in the United States, do you pay more attention, less attention, or the same attention to safety than you had to pay in your country?
- Which safety rules here are different from those in your country? What safety rules are the same?
- Are there safety committees in your country? If so, how often do they meet?
- In the companies you worked for in your country, were safety inspections conducted?
- From your work experience, what happened in case of a serious injury in your country?
- Did you use personal protective equipment in your country? Was it voluntary or mandatory?
- What are some of the reasons that people don't follow workplace safety rules even though they know them (whether in the United States or in your country)?

ESTABLISH A SAFETY AWARD PROGRAM

When an employee is observed assisting a co-worker to do the safe thing, that worker's name is put on the safety award list to receive a safety award at the next employee meeting.

HAZARD IDENTIFICATION CONTEST

A selection of many different prizes should be made available in the safety award program. The items don't need to be expensive but should be of value and usefulness. Here are some examples:

- Baseball hats
- Gift certificates
- Flashlights
- Compasses
- Golf clubs

You should have three gift certificates for the contest to stores in the community that carry a variety of desirable goods. Here are some examples:

- Sports stores
- Music and video stores
- Travel agencies
- Electronic stores
- Restaurants

It is suggested that the hazard identification contest have a higher award value than the safety award program. For example, if the gift certificates have values of $100 for first prize, $50 for second, and $25 for third, the total is only $175 per quarter or $700 per year . . . a small investment with a big safety payoff!

A supply of hazard identification sheets (see below) should be available that identify the hazard and the suggestion for correcting the hazard.

Hazard Identification

Date _____

Worker _____

Supervisor _____

Hazard _____

Hazard location _____

How to fix hazard _____

_____

_____

_____

Hazard Identification Contest Rules

Workers identify a hazard and a solution for that hazard. If they can't write, a co-worker or supervisor can assist in completing the form. The form is given to the supervisor. The supervisor immediately makes a copy of the form and gives the original to the person in charge of making repairs. The copy of the form is given to the chair of the safety committee.

Once per quarter, the safety committee takes the forms collected that quarter and elects the first, second, and third prizes based on rules established by the safety committee for this purpose. The award ceremony should be at a moment when all workers are gathered.

The hazard identification contest is one of the most powerful tools available to focus an entire workforce on solving workplace hazards. Not only is this contest an extraordinary morale booster but also it encourages workers to look for hazards! Considering that the average out-of-pocket cost to an employer for an accepted disabling injury is almost $10,000, the cost of the hazard identification contest is minimal.

Conduct Safety Training

The following tips will help train workers with limited English skills. Clearly identify the safety or health training topics.

- Demonstrate how to do the task safely.
- Workers should repeat the demonstration. Repeat the demonstrations until the task is completed correctly.
- Follow up after the end of the training program; if a worker is violating the safe way to do a task, immediately demonstrate the safe way to do the task. Have the worker

repeat the demonstration until the task is done safely. If the same worker violates the safe way to do a task after being effectively coached, initiate appropriate discipline with the worker.

People learn by listening, seeing, and touching. When someone does not understand the language, seeing and touching becomes more important than listening. So, train with visuals and hands-on teaching aids; this means pictures and objects!

The following tips will help identify the safety training topics:

- Clearly identify the topics with pictures and objects. For example, if the trainer is wearing gloves to handle chemicals, hold up the gloves, point to them, and say, "Today we are going to train you on gloves!"
- You may train on more than one topic if you clearly identify the beginning and end of each topic. For example, if in addition to training about gloves, the training session will also include reporting accidental chemical spills, say, "We are finished talking about gloves." Then, hold up a picture of a chemical spilling from a container or use a container as a prop and say, "Now we are going to train you on reporting chemical spills."

Demonstrate How to Do the Task Safely

The technical leader of the training team is probably the person for the task demonstrations! Jobs usually have several steps. Show each step by using actions. For example, operating a machine may require these five steps:

Step 1: Before starting, check and make sure that the machine is in good working condition. Let the workers see the trainer check the machine and explain what the trainer is checking.
Step 2: Verify that hazards are eliminated or controlled before starting the machine. Explain what the hazards are and how to eliminate or control those hazards.
Step 3: Demonstrate the correct method of using ear and eye protection. Show the correct method of inserting earplugs, then put on the eye protection.
Step 4: Start the machine, and verify it is running properly. Then explain how to verify that the machine is running properly.
Step 5: Operate the machine. Use the machine as intended by the manufacturer. Explain the correct use, the correct body position, and the correct method of shutting down the machine.

Volunteer Workers Need to Repeat the Demonstration

Repeat the demonstrations with volunteers until the task is completed correctly and safely. The following tips can help train workers with limited English skills.

- Observe the volunteers carefully as they repeat the demonstration.
- If the first volunteer does the job correctly the first time he or she attempts the task, congratulate him or her and ask another volunteer to repeat the demonstration.
- If a volunteer does not repeat the demonstration correctly, repeat the original demonstration. After repeating the original demonstration, ask the same volunteer to repeat the demonstration. Repeat this process until the first volunteer does the task correctly!
- Remember, it is your responsibility to ensure that the worker can do the job correctly and safely!

This famous anonymous quote applies when workers are asked to repeat the demonstration: "Give a man a fish; you have fed him for today. Teach a man to fish; you have fed him for a lifetime."

Follow-up

After the end of the training program, if a worker is violating the safe way to do a task, immediately repeat the demonstration. Have the worker repeat the demonstration until the task is done safely. If the same worker violates the safe way to do a task after being coached, initiate appropriate discipline with the worker.

There are several questions to ask before initiating a disciplinary process. If you can answer "yes" to each of these questions, discipline is appropriate:

- Have the necessary tools, equipment, and resources been provided?
- Have the workers been properly trained?
- Does the employee know he or she will be disciplined?
- Have the workers been supervised properly?

You are now ready to form a safety training team in your workplace!

CHAPTER 4

OSHA's Construction Industry Standards Simplified

Construction safety standards were initially adopted by the U.S. Department of Labor in April 1971 to implement the Contract Work Hours and Safety Standards Act, 40 USC 333, also known as the Construction Safety Act. Shortly thereafter, they were converted into Occupational Safety and Health Administration (OSHA) standards under the authority that Congress delegated to the secretary of labor in Section 6(a) of the Occupational Safety and Health Act (OSH Act).

An employer must comply with the safety and health regulations in Part 1926 if its employees are "engaged in construction work" (29 CFR 1910.12[b]). In order to determine if such work, defined as "work for construction, alteration, and/or repair, including painting and decorating," is being performed, one must consult 29 CFR 1926.13 for a discussion, which states the following:

> The terms "construction" . . . or "repair" mean all types of work done on a particular building or work at the site thereof . . . all work done in the construction or development of the project, including, without limitation, altering, remodeling, installation (where appropriate) on the site of the work of items fabricated off-site, painting and decorating, the transporting of materials and supplies to or from the building or work . . . manufacturing or furnishing of materials, articles, supplies or equipment on the site of the building or work.
>
> The site of work is limited to the physical place or places where the construction called for in the contract will remain when work on it has been completed. (29 CFR 5.2[l][1]).

The Part 1926 standards include 26 subparts (Subpart A through Subpart Z), but Subparts A and B apply only to determining the scope of Section 107 of the Construction Safety Act, 40 USC 333. That act applies only to employers who are engaged in construction under contract with the U.S. government. OSHA does not base citations upon either Subpart A or B. Consequently, no further consideration will be given to them in this book. The remaining 24 subparts are listed below:

- Subpart C   General safety and health provisions

- Subpart D   Occupational health and environmental controls
- Subpart E   Personal protective and life saving equipment
- Subpart F   Fire protection and prevention
- Subpart G   Signs, signals, and barricades
- Subpart H   Materials handling, storage, use, and disposal
- Subpart I   Tools—hand and power
- Subpart J   Welding and cutting
- Subpart K   Electrical
- Subpart L   Scaffolding
- Subpart M   Fall protection
- Subpart N   Cranes, derricks, hoists, elevators, and conveyors
- Subpart O   Motor vehicles, mechanized equipment, and marine operations
- Subpart P   Excavations
- Subpart Q   Concrete and masonry construction
- Subpart R   Steel erection
- Subpart S   Underground construction, caissons, cofferdams, and compressed air
- Subpart T   Demolition
- Subpart U   Blasting and the use of explosives
- Subpart V   Power transmission and distribution
- Subpart W   Rollover protective structures and overhead protection
- Subpart X   Stairways and ladders
- Subpart Y   Commercial diving operations
- Subpart Z   Toxic and hazardous substances

Employers engaged in construction should read over each of those subparts and identify those that could be applicable to their own operations, and then read the discussion of those subparts that follows.

OSHA has also identified a number of Part 1910 general industry standards that apply to construction work. In June 1993, each of them was assigned a Part 1926 designation and reissued as part of the construction standards. Thus, they will be included in the discussion that follows.

Subpart C—General Safety and Health Provisions

The general provisions included in §1926.20 require construction employers to

- initiate and maintain such programs as may be necessary to comply with the Part 1926 standards and to include as a requirement of those programs a provision requiring that frequent and regular inspections be made by "competent persons" of the job sites, materials, and equipment;
- identify (as unsafe) and prohibit the use of machinery, tools, materials, and equipment that are not in compliance with any Part 1926 standard; and
- limit the operation of equipment and machinery to only those employees who are qualified by training or experience to operate it.

Table 4.1.   Subpart C—General safety and health provisions

Section	Designation
1926.20	General safety and health provisions
1926.21	Safety training and education
1926.23	First aid and medical attention
1926.24	Fire protection and prevention
1926.25	Housekeeping
1926.26	Illumination
1926.27	Sanitation
1926.28	Personal protective equipment
1926.29	Acceptable certifications
1926.30	Shipbuilding and ship repairing
1926.31	Incorporation by reference
1926.32	Definitions
1926.33	Access to employee exposure and medical records
1926.34	Means of egress
1926.35	Employee emergency action plans

Some OSHA inspectors insist that §1926.20(b) requires that each construction employer must adopt a written accident-prevention program for each job site. The standard, however, does not include the word "written." Employers should pay close attention to §1926.20(b). It is one of the most cited construction safety standards.

Section 1926.21 requires that employees be given instruction (not training) in the recognition and avoidance of unsafe conditions and the regulations applicable (to his or her work) regarding hazard control, as well as the necessary precautions and the necessary protective equipment for those employees who are required to enter "confined or enclosed spaces." Although §1926.21 is somewhat ambiguous, it is a favorite of OSHA inspectors. Innumerable citations have been issued under that standard.

The standards in §§1926.22 through 1926.28 apply to most construction jobs and are stated succinctly and briefly. Each employer should familiarize himself or herself with them.

The "acceptable certifications" standard in §1926.29 applies only to boilers, pressure vessels, and similar equipment.

The provisions of §1926.30 and §1926.31 do not impose any substantive requirements on employers.

Section 1926.32 contains definitions for 18 terms that are used in Part 1926, including "competent person," "employee," "employer," "qualified," "safety factor," "shall," "should," and "suitable."

The three remaining standards are general industry standards that OSHA has incorporated into the construction standards. Section 1926.33 is the same as Section 1910.20. Section 1926.34 requires that there shall be free egress from every building and structure. It is essentially the same as Part 1910.

Section 1926.35 is the same as Section 1910.38, a standard that is also part of Part 1910, Subpart E—Means of egress.

Subpart D—Occupational Health and Environmental Controls

Table 4.2.   Subpart D—Occupational health and environmental controls

Section	Designation
1926.50	Medical services and first aid
1926.51	Sanitation
1926.52	Occupational noise exposure
1926.53	Ionizing radiation
1926.54	Nonionizing radiation
1926.55	Gases, vapors, fumes, dusts, and mists
1926.56	Illumination
1926.57	Ventilation
1926.59	Hazard communication
1926.60	Methylenedianiline (MDA)
1926.61	Retention of DOT markings, placards, and labels
1926.62	Lead
1926.64	Process safety management of highly hazardous chemicals
1926.65	Hazardous waste operations and emergency response
1926.66	Criteria for design and construction for spray booths

There are seven Part 1910 general industry standards that apply to Subpart D. They are as follows:

§1910.141(a)(1)   Scope of general industry sanitation standards
§1910.141(a)(2)(v)   "Potable water" defined
§1910.141(a)(5)   Vermin control
§1910.141(g)(2)   Eating and drinking areas
§1910.141(h)   Food handling
§1910.151(c)   Medical services and first aid—quick drenching facilities
§1910.161(a)(2)   Carbon dioxide extinguishing systems, safety requirements

The medical services, sanitation, and illumination requirements in §§1926.50, 1926.51, and 1926.56 apply at all construction sites.

Section 1926.52 is the noise standard for the construction industry. It limits employee noise exposure to 90 dBA averaged over an eight-hour day, but unlike the general industry noise standard, it does not require the same kind of hearing conservation program (monitoring, audiometric testing, etc.).

If you have employees exposed to high noise levels, you should also observe §1926.102 (part of Subpart E), which requires ear protective devices. For all intents and purposes, compliance with §1926.102 will satisfy the construction industry hearing conservation requirement.

The §1926.53 standard applies to activities that involve ionizing radiation, radioactive materials, or X-rays. The §1926.54 standard applies when laser equipment (nonionizing radiation) is used. The use of laser safety goggles is required by a Subpart E standard, §1926.102(b)(2).

Section 1926.55 is the construction equivalent of Part 1910, Subpart Z. As originally issued, it did not list the 400-odd regulated substances by name or the employee exposure limitations for those substances. It simply stated that the American Conference of Governmental Industrial Hygienists (ACGIH) Threshold Limit Values of Airborne Contaminants for 1970 shall be avoided. If not, controls must first be implemented whenever feasible. When not feasible, personal protective equipment must be used. Those ACGIH limits are now listed in appendix A of §1926.55. Construction employers should read the list of substances that appear in the standard's appendix A. If the limits listed there are exceeded, the §1926.55 requirements must be observed.

Section 1926.56 sets forth specific illumination intensities for construction areas, ramps, runways, corridors, offices, shops, and storage areas.

If ventilation is used as an engineering control to limit exposure to airborne contaminants, the §1926.57 Ventilation standard applies.

If your work involves exposure to Methylenedianiline (MDA), you must be familiar with §1926.60 and carefully observe its requirements. It includes innumerable details. This is also true for employers whose work involves exposure to lead. They must observe the numerous requirements included in the Lead standard (§1926.62).

Per §1926.61, each employer who receives a package of hazardous material that is required to be marked, labeled, or placarded in compliance with the Department of Transportation's (DOT) Hazardous Materials Regulations (49 CFR Parts 171 through 180) must keep those markings, labels, and placards on the package until the packaging is sufficiently cleaned of residue to remove any potential hazards. The markings, placards, and labels must be maintained in a manner that ensures that they are readily visible.

Process safety management of highly hazardous chemicals (§1926.64) is the same as the Part 1910 general industry standard.

Hazardous waste operations and emergency response (§1926.65) is also the same as the Part 1910 general industry standard. See the discussion later in this chapter under the heading "Subpart H—Materials Handling, Storage, Use, and Disposal."

Criteria for design and construction for spray booths (§1922.66) incorporates into the construction standards the general industry requirements (§1910.107) discussed later in this chapter under the heading "Subpart H—Materials Handling, Storage, Use, and Disposal."

The Hazard communication standard is the single most cited OSHA standard. Every employer must pay strict attention to §1926.59. It requires that chemical manufacturers and importers assess the hazards of all chemicals that they produce or import and furnish detailed material safety data sheets (MSDS) to their customers upon those determined to be hazardous. All employers must provide that information to their employees by means of a written hazard communication program, labels on containers, MSDSs, employee training, and access to written records of all of this.

The term "hazardous chemical" is defined very broadly, so virtually every employer is required to observe §1926.59.

Subpart E—Personal Protective and Life Saving Equipment

Table 4.3. Subpart E—Personal protective and life saving equipment

Section	Designation
1926.95	Criteria for personal protective equipment
1926.96	Occupational foot protection
1926.100	Head protection
1926.101	Hearing protection
1926.102	Eye and face protection
1926.103	Respiratory protection
1926.104	Safety belts, lifelines, and lanyards
1926.105	Safety nets
1926.106	Working over or near water
1926.107	Definitions applicable to this subpart

The Subpart E standards state in very general terms the conditions under which the protective equipment identified in the standards' titles must be used. Section 1926.95 is the same as Section 1910.132 discussed earlier in this book under the heading Subpart I—Personal Protective Equipment. Section 1926.96 simply sets forth the specifications for safety-toe footwear. Section 1926.105 discusses safety nets for elevated workplaces 25 feet or more above the adjoining surface if it is impractical to use ladders, scaffolds, temporary floors, or safety lines or belts as protection against falls. Section 1926.107 defines six of the terms used in Subpart E:

- Contaminant
- Lanyard
- Lifeline
- OD (optical density)
- Radiant energy
- Safety belt

Subpart F—Fire Protection and Prevention

Table 4.4. Subpart F—Fire protection and prevention

Section	Designation
1926.150	Fire protection
1926.151	Fire prevention
1926.152	Flammable and combustible liquids
1926.153	Liquefied petroleum gas (LP-Gas)
1926.154	Temporary heating devices
1926.155	Definitions applicable to this subpart

The fire protection and fire prevention requirements in §§1926.150 and 1926.151 are rather detailed and should be read and observed by all construction employers. Among other things, those standards require the development of a fire protection program (§1926.150[a]), a trained fire brigade "as warranted by the project" (§1926.150[a][5]), and periodic inspection and maintenance of all portable fire extinguishers (§1926.150[c][1][viii]).

If flammable or combustible liquids (or liquefied petroleum gas [LP-Gas]) are present at a construction site, the detailed requirements of §1926.152 (or §1926.153) must be observed.

Section 1926.154 set forth the rules for the use of temporary heating devices.

Subpart G—Signs, Signals, and Barricades

Table 4.5. Subpart G—Signs, signals, and barricades

Section	Designation
1926.200	Accident-prevention signs and tags
1926.201	Signaling
1926.202	Barricades
1926.203	Definitions applicable to this subpart

Subpart G is very brief. It requires the posting of construction areas with legible traffic signs at points of hazard (§1926.200[g]) and the use of flagmen or other appropriate traffic controls when the necessary protection cannot be provided by signs, signals, and barricades (§1926.201). The other Subpart G "standards" simply provide the specifications and color coding for signs and tags if they are used or required by some other OSHA standard.

Subpart H—Materials Handling, Storage, Use, and Disposal

Table 4.6. Subpart H—Materials handling, storage, use, and disposal

Section	Designation
1926.250	General requirements for storage
1926.251	Rigging equipment for material handling
1926.252	Disposal of waste materials

Section 1926.250 provides detailed requirements for the movement, stacking, and storage of various building material.

Section 1926.251 set forth the requirements to be observed whenever rigging equipment is used in material handling.

The disposal of ordinary waste materials (not hazardous waste) at a construction site is regulated by §1926.252.

Subpart I—Tools (Hand and Power)

Table 4.7. Subpart I—Tools (hand and power)

Section	Designation
1926.300	General requirements
1926.301	Hand tools
1926.302	Power-operated hand tools
1926.303	Abrasive wheels and tools
1926.304	Woodworking tools
1926.305	Jacks—lever and ratchet, screw and hydraulic
1926.306	Air receivers
1926.307	Mechanical power-transmission apparatus

The Subpart I standards set forth the requirements to be observed whenever the identified kinds of tools and equipment are used in construction. The General requirements standard, §1926.300, requires maintenance "in a safe condition" of all hand tools, power tools, and similar equipment; requires the guarding of moving parts of equipment; and imposes requirements for use of personal protective equipment (when using hand and power tools) and on-off controls on power tools.

Section 1926.306 is the same as Section 1910.169. See the discussion later in this chapter under "Subpart S—Underground Construction, Caissons, Cofferdams, and Compressed Air."

Section 1926.307 is the same as Section 1910.219. See the discussion later in this chapter under "Subpart O—Motor Vehicles, Mechanized Equipment, and Marine Operations." OSHA has also identified parts of four general industry standards that apply to Subpart I:

- Machine point of operations guarding (§1910.212[a][3] and §1910.212[a][5]).
- Anchoring fixed machinery (§1910.212[b]).
- Abrasive blast cleaning nozzles (§1910.244[b]).
- Jacks, operation and maintenance (§1910.244[a][2][iii] through §1910.244[a][2][viii]).

Subpart J—Welding and Cutting

Table 4.8. Subpart J—Welding and cutting

Section	Designation
1926.350	Gas welding and cutting
1926.351	Arc welding and cutting
1926.352	Fire prevention
1926.353	Ventilation and protection in welding, cutting, and heating
1926.354	Welding, cutting, and heating in way of preservative coatings

Subpart J regulates all aspects of welding, cutting, and heating when those operations are performed on a construction project. Employers should pay careful attention to those five Subpart J standards whenever welding or cutting is done.

Subpart K—Electrical

Table 4.9. Subpart K—Electrical

Division	Section	Designation
General	1926.400	Introduction
Installation safety requirements	1926.402	Applicability
	1926.403	General requirements
	1926.404	Wiring design and protection
	1926.405	Wiring methods, components, and equipment for general use
	1926.406	Specific purpose equipment and installations
	1926.407	Hazardous (classified) locations
	1926.408	Special systems
Safety-related work practices	1926.416	General requirements
	1926.417	Lockout and tagging of circuits
Safety-related maintenance and environmental considerations	1926.431	Maintenance of equipment
	1926.432	Environmental deterioration of equipment
Safety requirements for special equipment	1926.441	Battery locations and battery charging

Subpart K covers the electrical safety requirements for construction job sites. They are separated into four major divisions as indicated in table 4.9. The electrical standards are quite detailed, so the definitions in §1926.449 should be consulted when reading them.

Section 1926.400 simply states how the Electrical standards are arranged. Those responsible for construction job sites should familiarize themselves with the remaining provisions of Subpart K. If there are doubts or questions, the consultative service for your state is there to provide you with explanations.

Subpart L—Scaffolding

Table 4.10. Subpart L—Scaffolding

Section	Designation
1926.450	Scope and application
1926.451	General requirements
1926.452	Requirements for specific types of scaffolds
1926.453	Aerial lifts
1926.454	Training requirements

OSHA has revised the scaffolding standard to allow employers greater flexibility in the use of fall protection systems to protect employees working on scaffolds.

Employers who use scaffolds in construction work should be familiar with all the definitions included in §1926.450. Section 1926.451 contains all the general requirements for scaffolds. It also sets the minimum strength criteria for all scaffold components and connections. It requires that each scaffold component be capable of supporting, without failure, its own weight and at least four times the maximum intended load applied to it. Section 1926.452 has specific requirements for 25 particular types of scaffolds (listed in table 4.11).

Table 4.11. Specific requirements for scaffolds

Scaffold component	Section
Pole	§1926.452(a)
Tube and coupler	§1926.452(b)
Fabricated frame	§1926.452(c)
Plasterers, decorators, and large area scaffolds	§1926.452(d)
Bricklayers square	§1926.452(e)
Horse	§1926.452(f)
Form and carpenters	§1926.452(g)
Roof bracket	§1926.452(h)
Outrigger	§1926.452(i)
Pump jack	§1926.452(j)
Ladder jack	§1926.452(k)
Window jack chairs	§1926.452(l)
Crawling boards	§1926.452(m)
Step platform and trestle ladder	§1926.452(n)
Single-point adjustable	§1926.452(o)
Two-point adjustable suspension (swing stages)	§1926.452(p)
Multipoint suspension	§1926.452(q)
Catenary	§1926.452(r)
Float or ship	§1926.452(s)
Interior hung	§1926.452(t)
Needle beam	§1926.452(u)
Multilevel	§1926.452(v)
Mobile	§1926.452(w)
Repair bracket	§1926.452(x)
Stilts	§1926.452(y)

Section 1926.453 is a recent addition to the scaffold standard. It sets some specific requirements for lift operations. It applies to equipment for vehicle-mounted elevating and rotating work platforms.

Section 1926.454 covers training requirements. The standard distinguishes between the training needed by employees to erect and to dismantle scaffolds. The standard applies to all construction work. It requires employers to instruct each employee in the recognition and avoidance of unsafe conditions. It also sets certain criteria allowing employers to tailor training to fit the particular circumstances of each employer's workplace.

Subpart M—Fall Protection

Table 4.12. Subpart M—Fall protection

Section	Designation
1926.500	Scope, application, and definitions applicable to this subpart
1926.501	Duty to have fall protection
1926.502	Fall protection systems
1926.503	Training requirements

The Fall protection standard contains OSHA's revised regulations for construction workplaces to prevent employees from falling off, onto, or through levels, and to protect employees from being struck by falling objects.

Under the standard, employers will be able to select fall protection measures compatible with the type of work being performed. Fall protection generally can be provided through the use of guardrail systems, safety net systems, personal fall arrest systems, positioning device systems, and warning line systems, among others.

Section 1926.501 requires employers to assess the workplace to determine if the walking/working surfaces on which employees are to work have the strength and structural integrity to safely support workers. Employees are not permitted to work on those surfaces until it has been determined that the surfaces have the requisite strength and structural integrity to support the workers. Once employers have determined that the surface is safe for employees to work on, the employer must select one of the options listed for the work operation if a fall hazard is present.

For example, if an employee is exposed to falling six feet or more from an unprotected side or edge, the employer must select either a guardrail system, safety net system, or personal fall arrest system to protect the worker.

Under §1926.502, fall protection can be provided through the use of guardrail systems, safety net systems, or personal fall arrest systems. However, under certain conditions, employers that cannot use conventional fall protection equipment must develop and implement a written fall protection plan specifying alternative fall protection measures.

If the employer chooses to use guardrail systems to protect workers from falls, the system must meet the following criteria: guardrails require a toprail at around 42 inches, a midrail at around 21 inches, and toeboards when necessary to prevent objects from falling over the edge.

If an employer chooses to use safety net systems, the system must be installed as close as practicable under the walking/working surface on which employees are working and never more than 30 feet (9.1 meters) below such levels. Defective nets cannot be used. Safety nets must be inspected at least once a week for wear, damage, and other deterioration. The maximum size of each safety net mesh opening is neither to exceed 36 square inches (230 square centimeters) nor be longer than six inches (15 centimeters) on any side, and the openings, measured center to center, of mesh ropes or webbing shall not exceed six inches (15 centimeters). All mesh crossings shall be secured to prevent enlargement of the mesh opening.

Safety nets must be capable of absorbing an impact force of a drop test consisting of a 400-pound (180 kilogram) bag of sand 30 inches (76 centimeters) in diameter dropped from the highest walking/working surface at which workers are exposed but not from less than 42 inches (1.1 meters) above that level.

If an employer chooses to use a personal fall arrest system, it must consist of an anchorage, connectors, and a body belt or body harness and might include a deceleration device, lifeline, or suitable combinations. If a personal fall arrest system is used for fall protection, it must do the following:

- Limit maximum arresting force on an employee to 900 pounds (four kilonewtons) when used with a body belt.

- Limit maximum arresting force on an employee to 1,800 pounds (eight kilonewtons) when used with a body harness.
- Be rigged so that employees can neither free-fall more than six feet (1.8 meters) nor contact any lower level.
- Bring an employee to a complete stop and limit maximum deceleration distance an employee travels to 3.5 feet (1.07 meters).
- Have sufficient strength to withstand twice the potential impact energy of an employee free-falling a distance of six feet (1.8 meters) or the free-fall distance permitted by the system, whichever is less.

The use of body belts for fall arrest is no longer allowed, effective January 1, 1998.

Personal fall arrest systems must be inspected prior to each use for wear damage and other deterioration. Defective components must be removed from service. Dee-rings and snaphooks must have a minimum tensile strength of 5,000 pounds (22.2 kilonewtons). Dee-rings and snaphooks must be proof tested to a minimum tensile load of 3,600 pounds (16 kilonewtons) without cracking, breaking, or suffering permanent deformation.

Snaphooks must be sized to be compatible with the member to which they will be connected or be of a locking configuration.

Horizontal lifelines must be installed and used under the supervision of a qualified person. Self-retracting lifelines and lanyards that automatically limit free-fall distance to two feet (0.61 meters) or less must be capable of sustaining a minimum tensile load of 3,000 pounds (13.3 kilonewtons) applied to the device with the lifeline or lanyard in the fully extended position.

Self-retracting lifelines and lanyards that do not limit free-fall distance to two feet (0.61 meters) or less, ripstitch lanyards, and tearing and deforming lanyards must be capable of sustaining a minimum tensile load of 5,000 pounds (22.2 kilonewtons) applied to the device with the lifeline or lanyard in the fully extended position.

Anchorages must be designed, installed, and used under the supervision of a qualified person, as part of a complete personal fall arrest system that maintains a safety factor of at least two, that is, capable of supporting at least twice the weight expected to be imposed upon it. Anchorages used to attach personal fall arrest systems must be independent of any anchorage being used to support or suspend platforms and must be capable of supporting at least 5,000 pounds (22.2 kilonewtons) per person attached.

Lanyards and vertical lifelines must have a minimum breaking strength of 5,000 pounds (22.2 kilonewtons).

Other acceptable fall protection systems under §1926.502 can be used, such as controlled access zones that are allowed as part of a fall protection plan when designed according to OSHA requirements. Safety monitoring systems will also be acceptable as part of a fall protection plan. Covers will be utilized to protect against fall hazards (like floor holes).

Written fall protection plans may be used when employers can demonstrate that conventional fall protection systems pose a greater hazard to employees than using alternative systems.

A warning line system can also be used. It can consist of ropes, wires, or chains. Warning lines must be erected around all sides of roof work areas. When mechanical equipment is being used, the warning line must be erected not less than six feet (1.8 meters) from the roof edge parallel to the direction of mechanical equipment operation and not less than 10 feet (3 meters) from the roof edge perpendicular to the direction of mechanical equipment operation.

Section 1926.503 covers training requirements. Employers must provide a training program that teaches employees who might be exposed to fall hazards how to recognize such hazards and how to minimize them. Employees must be trained in the following areas:

- The nature of fall hazards in the work area;
- The correct procedures for erecting, maintaining, disassembling, and inspecting fall protection systems;
- The use and operation of controlled access zones and guardrail, personal fall arrest, safety net, warning line, and safety monitoring systems;
- The role of each employee in safety monitoring when the system is in use;
- The limitations on the use of mechanical equipment during the performance of roofing work on low-sloped roofs;
- The correct procedures for equipment and materials handling and storage and the erection of overhead protection; and
- The employees' role in fall protection plans.

The employer must prepare a written certification that identifies the employee trained and the date of the training. The employer or trainer must sign the certification record, and retraining must be provided when necessary.

Subpart N—Cranes, Derricks, Hoists, Elevators, and Conveyors

Table 4.13. Subpart N—Cranes, derricks, hoists, elevators, and conveyors

Section	Designation
1926.550	Cranes and derricks
1926.551	Helicopter cranes
1926.552	Material hoists, personnel hoists, and elevators
1926.553	Base-mounted drum hoists
1926.554	Overhead hoists
1926.555	Conveyors

Construction employers who use the kinds of lifting equipment and mechanisms identified in the titles of these six Subpart N standards must familiarize themselves with the applicable requirements and observe them.

Subpart O—Motor Vehicles, Mechanized Equipment, and Marine Operations

Table 4.14.   Subpart O—Motor vehicles, mechanized equipment, and marine operations

Section	Designation
1926.600	Equipment
1926.601	Motor vehicles
1926.602	Material handling equipment
1926.603	Pile driving equipment
1926.604	Site clearing
1926.605	Marine operations and equipment
1926.606	Definitions applicable to this subpart

The Subpart O standards apply to the equipment and operations listed in the titles of its six substantive standards.

Section 1926.600 imposes various requirements for (1) equipment that is left unattended at night; (2) when employees inflate, mount, or dismount tires installed on split rims or rims equipped with locking rims or similar devices; (3) heavy equipment or machinery that is parked, suspended, or held aloft by slings, hoists, or jacks; 4) care and charging batteries; (5) when the equipment is being moved in the vicinity of power lines or energized transmitters; and (6) when rolling cars are on spur railroad tracks. It also requires that cab glass must be safety glass (or equivalent) so that there will be no visible distortion to affect the safe operation of the machines that are regulated in Subpart O.

Section 1926.601 has various requirements for vehicles that operate within an off-highway job site that is not open to public traffic.

Section 1926.602 applies to earthmoving equipment, off-highway trucks, rollers, compactors, front-end loaders, bulldozers, tractors, lift trucks, stackers, high-lift rider industrial trucks, and similar construction equipment.

Pile driving and the equipment used in that operation is regulated by §1926.603.

Section 1926.604 requires rollover guards and canopy guards on equipment used in site-clearing operations and provides that employees engaged in such operations be protected from irritant and toxic plants and instructed in the first aid treatment that is available.

Section 1926.605 has various requirements upon employers engaged in marine construction operations.

OSHA rules for other vehicles frequently used in construction are included in Subpart W. See the discussion of that subpart later in this chapter.

Subpart P—Excavations

Table 4.15.   Subpart P—Excavations

Section	Designation
1926.650	Scope, application, and definitions applicable to this subpart
1926.651	Specific excavation requirements
1926.652	Requirements for protective systems

The Subpart P standards apply to all open excavations (including trenches) made in the earth's surface. In some situations, it requires the use of written designs (approved by a registered professional engineer) for sloping, benching, and support systems.

There are six explanatory appendixes to Subpart P that should be consulted by employers engaged in excavation work. They include illustrations, tables, and diagrams. They cover the following subjects:

Appendix A   Soil classification
Appendix B   Sloping and benching
Appendix C   Timber shoring for trenches
Appendix D   Aluminum hydraulic shoring for trenches
Appendix E   Alternatives to timber shoring
Appendix F   Selection of protective systems

The Subpart P standards went into effect in 1990. They constitute a total revision of the OSHA trenching and excavation standards that had been in effect for the previous 18 years.

Section 1926.651 has specific requirements for excavations that need to be reviewed for further details. Daily inspections of excavations must be made by a competent person to make sure that cave-ins, failure of protective systems, or hazardous atmospheres are not found. If the competent person discovers there are indications that could result in a possible cave-in, exposed employees must be removed from the hazardous area until the necessary precautions have been taken.

Subpart Q—Concrete and Masonry Construction

Table 4.16.   Subpart Q—Concrete and masonry construction

Section	Designation
1926.700	Scope, application, and definitions applicable to this subpart
1926.701	General requirements
1926.702	Requirements for equipment and tools
1926.703	Requirements for cast-in-place concrete
1926.704	Requirements for precast concrete
1926.705	Requirements for lift-slab construction operations
1926.706	Requirements for masonry construction

Subpart Q includes detailed requirements to be observed in concrete and masonry construction operations.

A half-dozen general rules that apply to all such operations are set forth in §1926.701.

Section 1926.702 includes requirements for bulk storage bins, containers, and silos (§1926.702[a]); concrete mixers (§1926.702[b]); powered and rotating-type troweling machines (§1926.702[c]); concrete buggies (§1926.702[d]); concrete pumping systems (§1926.702[e]); concrete buckets (§1926.702[f]); tremies (§1926.702[g]); bull floats (§1926.702[h]); masonry saws (§1926.702[i]); and lockout/tagout procedures for compressors, mixers, and screens or pumps that are used in concrete and masonry construction activities (§1926.702[j]).

Detailed requirements for three specialized types of operations (cast-in-place concrete, precast concrete, and lift-slab construction) are included in §1926.703 through §1926.705. Brief explanatory appendixes are provided in both §1926.703 and §1926.705. The particular requirements for masonry construction are set forth in §1926.706.

Part 3 at the end of Subpart Q (chapter 10) lists 17 reference sources that can be helpful in understanding its various requirements.

Subpart R—Steel Erection

Table 4.17. Subpart R—Steel erection

Section	Designation
1926.750	Scope
1926.751	Definitions
1926.752	Site layout, site-specific erection plan, and construction sequence
1926.753	Hoisting and rigging
1926.754	Structural steel assembly
1926.755	Column anchorage
1926.756	Beams and columns
1926.757	Open web steel joists
1926.758	Systems-engineered metal buildings
1926.759	Falling object protection
1926.760	Fall protection
1926.761	Training

The steel erection standard enhances protections provided to ironworkers by addressing the hazards that have been identified as the major causes of injuries and fatalities in the steel erection industry. These are hazards associated with working under loads; hoisting, landing, and placing decking; column stability; double connections; landing and placing steel joists; and falls to lower levels.

Key provisions of the revised steel erection standard include the following:

Site layout and construction sequence
- Requires certification of proper curing of concrete in footings, piers, and so forth, for steel columns; and
- Requires controlling contractor to provide erector with a safe site layout including preplanning routes for hoisting loads.

Site-specific erection plan
- Requires preplanning of key erection elements, including coordination with controlling contractor before erection begins, in certain circumstances.

Hoisting and rigging
- Provides additional crane safety for steel erection;
- Minimizes employee exposure to overhead loads through preplanning and work practice requirements; and

- Prescribes proper procedure for multiple lifts (Christmas-treeing).

Structural steel assembly

- Provides safer walking/working surfaces by eliminating tripping hazards and minimizes slips through new slip resistance requirements; and
- Provides specific work practices regarding safely landing deck bundles and promoting the prompt protection from fall hazards in interior openings.

Column anchorage

- Requires four anchor bolts per column along with other column stability requirements; and
- Requires procedures for adequacy of anchor bolts that have been modified in the field.

Beams and columns

- Eliminates extremely dangerous collapse hazards associated with making double connections at columns.

Open web steel joists

- Requires minimizing collapse of lightweight steel joists by addressing need for erection bridging and method of attachment;
- Requires bridging terminus anchors with illustrations and drawings in a nonmandatory appendix (provided by the Steel Joist Industry); and
- Newly requires minimizing collapse in placing loads on steel joists.

Systems-engineered metal buildings

- Requires minimizing collapse in the erection of these specialized structures that account for a major portion of steel erection in this country.

Falling object protection

- Provides performance provisions that address hazards of falling objects in steel erection.

Fall protection

- Provides controlled decking zone (CDZ) provisions to prevent decking fatalities;
- Requires that deckers in a CDZ and connectors be protected at heights greater than two stories or 30 feet and that connectors between 15 and 30 feet wear fall arrest or restraint equipment and be able to be tied off or be provided another means of fall protection; and
- Requires fall protection for all others engaged in steel erection at heights greater than 15 feet.

Training

- Requires a qualified person to train exposed workers in fall protection; and
- Requires a qualified person to train exposed workers engaged in special, high-risk activities.

Subpart S—Underground Construction, Caissons, Cofferdams, and Compressed Air

Subpart S applies only to underground construction work including tunnels, shafts, chambers and passageways; caisson work; work on cofferdams; and work conducted in a compressed air environment. Decompression tables are attached to Subpart S as chapter 10. The four standards that make up Subpart S are very detailed, so employers who do that kind of work should pay strict attention to them.

Table 4.18. Subpart S—Underground construction, caissons, cofferdams, and compressed air

Section	Designation
1926.800	Underground construction
1926.801	Caissons
1926.802	Cofferdams
1926.803	Compressed air
1926.804	Definitions applicable to this subpart

Subpart T—Demolition

Table 4.19. Subpart T—Demolition

Section	Designation
1926.850	Preparatory operations
1926.851	Stairs, passageways, and ladders
1926.852	Chutes
1926.853	Removal of materials through floor openings
1926.854	Removal of walls, masonry sections, and chimneys
1926.855	Manual removal of floors
1926.856	Removal of walls, floors, and material with equipment
1926.857	Storage
1926.858	Removal of steel construction
1926.859	Mechanical demolition
1926.860	Selective demolition by explosives

The Subpart T standards are restricted in their application to employers engaged in demolition operations. They are neither lengthy nor complicated. Employers engaged in those operations should have little difficulty understanding them.

Subpart U—Blasting and Use of Explosives

Table 4.20. Subpart U—Blasting and use of explosives

Section	Designation
1926.900	General provisions
1926.901	Blaster qualifications
1926.902	Surface transportation of explosives
1926.903	Underground transportation of explosives
1926.904	Storage of explosives and blasting agents
1926.905	Loading of explosives or blasting agents
1926.906	Initiation of explosive charges—electrical blasting
1926.907	Use of safety fuse
1926.908	Use of detonating cord
1926.909	Firing the blast
1926.910	Inspection after blasting
1926.911	Misfires
1926.912	Underwater blasting
1926.913	Blasting in excavation work under compressed air
1926.914	Definitions applicable to this subpart

Subpart U applies when employers use explosives or do blasting in their work. No employer should use explosives unless he or she is familiar with the Subpart U standards. They impose detailed restrictions on those working with explosives.

OSHA has identified parts of four general industry standards that apply to Subpart U:

- Buildings used for blasting agent mixing (§1910.109[g][2][ii])
- Buildings used for the mixing of water gels (§1910.109[h][3][ii])
- Semiconductive hose for explosives and blasting agents (§1910.109[a][12])
- Pneumatic blasting agent loading over blasting caps (§1910.109[e][3][iii])

Subpart V—Power Transmission and Distribution

Table 4.21. Subpart V—Power transmission and distribution

Section	Designation
1926.950	General requirements
1926.951	Tools and protective equipment
1926.952	Mechanical equipment
1926.953	Material handling
1926.954	Grounding for protection of employees
1926.955	Overhead lines
1926.956	Underground lines
1926.957	Construction in energized substations
1926.958	External load helicopters
1926.959	Lineman's body belts, safety straps, and lanyards
1926.960	Definitions applicable to this subpart

The Subpart V standards are limited in their application to construction, erection, alteration, conversion, and improvement of electric transmission and distribution lines and equipment. Employers engaged in that type of work must familiarize themselves with those standards. Employers who do not do work of that kind will have no reason to review the Subpart V standards.

Subpart W—Rollover Protective Structures; Overhead Protection

Table 4.22. Subpart W—Rollover protective structures; overhead protection

Section	Designation
1926.1000	Rollover protective structures (ROPS) for material handling equipment
1926.1001	Minimum performance criteria for rollover protective structures for designated scrapers, loaders, dozers, graders, and crawler tractors
1926.1002	Protective frame (ROPS) test procedures and performance requirements for wheel-type agricultural and industrial tractors used in construction
1926.1003	Overhead protection for operators of agricultural and industrial tractors

Subpart W only applies to seven kinds of material handling equipment when used in construction work: (1) rubber-tired, self-propelled scrapers; (2) rubber-tired front-end loaders; (3) rubber-tired dozers; (4) wheel-type agricultural and industrial tractors; (5) crawler tractors; (6) crawler-type loaders; and (7) motor graders, with or without attachments.

Employers who use any of the seven kinds of equipment listed above should familiarize themselves with the applicable requirements. Those who don't use that equipment should have no reason to be concerned with Subpart W.

Subpart X—Stairways and Ladders

Table 4.23. Subpart X—Stairways and ladders

Section	Designation
1926.1050	Scope, application, and definitions applicable to this subpart
1926.1051	General requirements
1926.1052	Stairways
1926.1053	Ladders
1926.1060	Training requirements

The Subpart X standards went into effect in 1991. They set forth the conditions and circumstances under which ladders and stairways must be provided (and used). The standards apply to all stairways and ladders used in construction, alteration, repair, demolition, painting, and repair work (however, some of the requirements do not apply to some ladders and stairways that were built prior to March 15, 1991).

Any employer who has one or more employees engaged in construction work who will use a ladder or stairway while working (and that covers the vast majority of construction employers) must become familiar with the Subpart X requirements.

Section 1926.1060 requires that a training program be provided for each employee using ladders and stairways so that he or she will be able to recognize the hazards related to ladders and stairways and know the procedures to be followed in order to minimize those hazards.

An appendix (appendix A) has been added to Subpart X to assist employers in complying with the loading and strength requirements for ladders that are set forth in §1926.1053(a)(1).

Subpart Z—Toxic and Hazardous Substances

Subpart Z contains 24 substance-specific standards that are identical to the general industry standards (1910) that regulate those same substances. Those standards are designated in the construction industry standards.

New Hexavalent Chromium Standard

On February 28, 2006, OSHA published the final Hexavalent Chromium (Cr[VI]) Standard Construction (1926.1126). The new standard lowers OSHA's permissible exposure limit (PEL) for hexavalent chromium, and for all Cr(VI) compounds, from 52 to 5 micrograms of Cr(VI) per cubic meter of air as an eight-hour time-weighted average (TWA). The standard also includes provisions relating to preferred methods for controlling exposure, respiratory protection, protective work clothing and equipment, hygiene areas and practices, medical surveillance, hazard communication, and recordkeeping.

The respiratory protection requirements for the three standards are similar. A respiratory protection program, including respirator selection, is required to follow OSHA Section 1910.134.

Employers must reassess their exposure controls, including the adequacy of their respirator program, taking into consideration the lower exposure limit.

If they have not done so already, employers in the affected industries should make an exposure determination to establish whether or not the new standard and its requirements apply and, if so, implement the necessary steps for compliance, including selection of proper respirators.

POTENTIAL HEALTH EFFECTS

The primary health impairments from workplace exposure to Cr(VI) are lung cancer, asthma, and damage to the nasal epithelia and skin. For industrial exposure, inhalation and the skin are the primary routes of uptake. Occupational exposure to Cr(VI) can lead to nasal tissue ulcerations and nasal septum perforations. Effects on the skin are the result of two distinct processes:

- Irritant reactions, such as skin ulcers and irritant contact dermatitis, and
- Delayed hypersensitivity (allergic) reactions.

ARE YOUR WORKERS EXPOSED TO HEXAVALENT CHROMIUM?

This standard applies to all occupational exposures to Cr(VI) in construction except the following:

- Exposures that occur in the application of pesticides regulated by the Environmental Protection Agency or another federal government agency (e.g., the treatment of wood with preservatives);
- Exposures to Portland cement; or
- Where the employer has objective data demonstrating that a material containing chromium or a specific process, operation, or activity involving chromium cannot release dust, fumes, or mists of chromium (VI) in concentrations at or above 0.5µg/m3 as an eight-hour time-weighted average under any expected conditions of use.

PERMISSIBLE EXPOSURE LIMIT (PEL)

The employer shall ensure that no employee is exposed to an airborne concentration of chromium (VI) in excess of 5 micrograms per cubic meter of air (5μg/m3), calculated as an eight-hour time-weighted average.

The permissible exposure limit (PEL) for all industries is 5μg/m3. The action level, or the level where requirements of the standard such as medical surveillance may be required, is 2.5μg/m3. There is no short-term exposure limit (STEL).

EXPOSURE DETERMINATION

Each employer who has a workplace or work operation shall determine the eight-hour TWA exposure for each employee exposed to chromium (VI). Two options are permitted. To select the proper respirator, the employer must make an exposure determination. The standard permits this to be done in one of two ways:

1. The "scheduled monitoring option" requires air sampling to make an initial characterization of worker exposures. The employer shall perform initial monitoring to determine the eight-hour TWA exposure for each employee on the basis of a sufficient number of personal breathing zone air samples to accurately characterize full shift exposure on each shift, for each job classification, and in each work area.

   Depending on the exposures found, sampling may need to be repeated every three months or every six months. Monitoring methods for hexavalent chromium include the National Institute for Occupational Safety and Health (NIOSH) Methods 7604 (by ion chromatography) and 7600 (by visible absorption spectrophotometry) or OSHA Method ID-215 (noted in the hexavalent chromium standard). Air sampling performed to comply with either option must have accuracy of ± 25 percent at the 95 percent confidence interval. Consult an American Industrial Hygiene Association (AIHA) accredited laboratory for assistance on selection of the appropriate sampling and analytical method.
2. Alternatively, OSHA allows the "performance-oriented option" where exposures can be estimated using any combination of air sampling, historical monitoring data, or objective data. Objective data means information such as air monitoring data from industrywide surveys or calculations based on the composition or chemical and physical properties of a substance demonstrating the employee exposure to CrVI associated with a particular product or material or a specific process, operation, or activity. The data must reflect workplace conditions closely resembling the processes, types of material, control methods, work practices, and environmental conditions in the employer's current operations.

METHODS OF COMPLIANCE

The employer shall use engineering and work practice controls to reduce and maintain employee exposure to chromium (VI) to or below the PEL unless the employer can demonstrate that such controls are not feasible.

Wherever feasible engineering and work practice controls are not sufficient to reduce employee exposure to or below the PEL, the employer shall use them to reduce employee exposure to the lowest levels achievable and shall supplement them by the use of respiratory protection that complies with the requirements of Section 1910.134.

EXCEPTIONS

Where painting of aircraft or large aircraft parts is performed in the aerospace industry, the employer shall use engineering and work practice controls to reduce and maintain employee exposure to chromium (VI) to or below 25µg/m3 unless the employer can demonstrate that such controls are not feasible. The employer shall supplement such engineering and work practice controls with the use of respiratory protection that complies with the requirements of Section 1910.134 to achieve the PEL of 5µg/m3.

Where the employer can demonstrate that a process or task does not result in any employee exposure to chromium (VI) above the PEL for 30 or more days per year (12 consecutive months), the requirement to implement engineering and work practice controls to achieve the PEL does not apply to that process or task. Rotation of employees to different jobs to comply with the PEL is not permitted.

RESPIRATORY PROTECTION

The employer shall provide respiratory protection for employees in the following circumstances:

- While engineering and work practice controls are being developed;
- During maintenance and repair activities for which engineering and work practice controls are not feasible;
- When all feasible engineering and work practice controls are implemented and are still not sufficient to reduce exposures to or below the PEL; and
- When employees are exposed above the PEL for fewer than 30 days per year, and the employer has elected not to implement engineering and work practice controls.

In Emergencies where respirator use is required, the employer shall institute a respiratory protection program in accordance with 29 CFR 1910.134. This applies to all aspects of respirator selection use and care. Thus, as with most other air contaminants, elastomeric or filtering half-face pieces with class 95 filters may be used up to 10 times the PEL, and full-face pieces with class 95 filters may be used up to 50 times the PEL when quantitatively fit tested, and so forth. If oil is present in the atmosphere along with Cr(VI), R or P series filters must be used.

PROTECTIVE WORK CLOTHING AND EQUIPMENT

Where a hazard is present or is likely to be present from skin or eye contact with chromium (VI), the employer shall provide appropriate personal protective clothing and equipment at no cost to employees and shall ensure that employees use it.

MEDICAL SURVEILLANCE

The employer shall make medical surveillance available at no cost to the employee, and at a reasonable time and place, for all employees

- who are or might be occupationally exposed to chromium (VI) at or above the action level for 30 or more days a year; or
- who are experiencing signs or symptoms of the adverse health effects associated with chromium (VI) exposure.

All medical examinations and procedures must be performed by or under the supervision of a Physician or Other Licensed Health Care Professional (PLHCP). The employer shall provide the PLHCP with a copy of this standard; a description of the affected employee's former, current, and anticipated duties as they relate to the employee's occupational exposure to chromium (VI); the employee's former, current, and anticipated levels of occupational exposure to chromium (VI); and a description of any personal protective equipment used or to be used by the employee, including when and for how long the employee has used that equipment.

The employer shall obtain a written medical opinion from the PLHCP that contains, among other things, any recommended limitations upon the employee's exposure to chromium (VI) or upon the use of personal protective equipment such as respirators.

COMMUNICATION OF CHROMIUM (VI) HAZARDS TO EMPLOYEES

In addition to the requirements of the Hazard communication standard, 29 CFR 1926.59, employers shall ensure that each employee can at least demonstrate knowledge of the contents of this section and the purpose and a description of the required medical surveillance program.

RECORDKEEPING

The employer shall maintain an accurate record of all air monitoring conducted to comply with the requirements of this section. This record shall include at least the following information:

- The date of measurement for each sample taken;
- The operation involving exposure to chromium (VI) that is being monitored;
- Sampling and analytical methods used and evidence of their accuracy;

- Number, duration, and the results of samples taken;
- Type of personal protective equipment, such as respirators, worn; and
- Name, social security number, and job classification of all employees represented by the monitoring, indicating which employees were actually monitored.

HISTORICAL MONITORING DATA

Where the employer has relied on historical monitoring data to determine exposure to chromium (VI), the employer shall establish and maintain an accurate record of the historical monitoring data relied upon. The record shall include information that reflects the following conditions:

- The data were collected using methods that meet the accuracy requirements;
- The processes and work practices that were in use when the historical monitoring data were obtained are essentially the same as those to be used during the job for which exposure is being determined;
- The characteristics of the chromium-containing material being handled when the historical monitoring data were obtained are the same as those on the job for which exposure is being determined; and
- Environmental conditions prevailing when the historical monitoring data were obtained are the same as those on the job for which exposure is being determined.

OBJECTIVE DATA

The employer shall maintain an accurate record of all objective data relied upon to comply with the requirements of this section. This record shall include at least the following information:

- The chromium-containing material in question;
- The source of the objective data;
- The testing protocol and results of testing, or analysis of the material for the release of chromium (VI);
- A description of the process, operation, or activity and how the data support the determination; and
- Other data relevant to the process, operation, activity, material, or employee exposures.

MEDICAL SURVEILLANCE

The employer shall establish and maintain an accurate record for each employee covered by medical surveillance. The record shall include the following information about the employee:

- Name and social security number;
- A copy of the PLHCP's written opinions; and
- A copy of the information provided to the PLHCP.

All medical records must be maintained and made available in accordance with 29 CFR 1910.1020.

Table 4.24. Subpart Z—Toxic and hazardous substances

Subpart Z—Toxic and hazardous substances	
§1926.1101	Asbestos
§1926.1102	Coal tar pitch volatiles: interpretation of term
§1926.1103	13 carcinogens (4-Nitrobiophenyl, etc.)
§1926.1104	alpha-Naphthylamine
§1926.1106	Methyl chloromethyl ether
§1926.1107	3.3'-Dichlorobenzidine (and its salts)
§1926.1108	bis-Chloromethyl ether
§1926.1109	beta-Naphthylamine
§1926.1110	Benzidine
§1926.1111	4-Aminodiphenyl
§1926.1112	Ethyleneimine
§1926.1113	beta-Propiolactone
§1926.1114	2-Acetylaminofluorene
§1926.1115	4-Dimethylaminoazobenzene
§1926.1116	N-Nitrosodimethylamine
§1926.1117	Vinyl chloride
§1926.1118	Inorganic arsenic
§1926.1126	Hexavalent chromium, Cr(VI)
§1926.1127	Cadmium
§1926.1128	Benzene
§1926.1129	Coke oven emissions
§1926.1144	1,2-dibromo-3-chloropropane
§1926.1145	Acrylonitrile
§1926.1147	Ethylene oxide
§1926.1148	Formaldehyde
§1926.1152	Methylene chloride

CHAPTER 5

OSHA's Construction Training Standards Simplified

Industry training standards are located in 29 CFR 1926. There is only one volume in the Code of Federal Regulations that contains them. Employers engaged in construction should read over each of those subparts and identify those that could be applicable to their own operations, and then read the discussion of those subparts that follow. You don't have to read them all. Some will obviously not be applicable to your business. Forget about them. Concentrate on the ones that might be applicable.

The Part 1926 standards include 26 subparts (Subpart A through Subpart Z), but Subparts A and B apply only to determining the scope of Section 107 of the Construction Safety Act, 40 USC 333. That act applies only to employers who are engaged in construction under contract with the U.S. government. The Occupational Safety and Health Administration (OSHA) does not base citations upon either Subpart A or B. Consequently, no further consideration will be given to them in this book. The remaining 24 subparts are listed below:

- Subpart C—General safety and health provisions
- Subpart D—Occupational health and environmental controls
- Subpart E—Personal protective and life saving equipment
- Subpart F—Fire protection and prevention
- Subpart G—Signs, signals, and barricades
- Subpart H—Materials handling, storage, use, and disposal
- Subpart I—Tools (hand and power)
- Subpart J—Welding and cutting
- Subpart K—Electrical
- Subpart L—Scaffolding
- Subpart M—Fall protection
- Subpart N—Cranes, derricks, hoists, elevators, and conveyors
- Subpart O—Motor vehicles, mechanized equipment, and marine operations
- Subpart P—Excavations
- Subpart Q—Concrete and masonry construction
- Subpart R—Steel erection

- Subpart S—Underground construction, caissons, cofferdams, and compressed air
- Subpart T—Demolition
- Subpart U—Blasting and the use of explosives
- Subpart V—Power transmission and distribution
- Subpart W—Rollover protective structures; overhead protection
- Subpart X—Stairways and ladders
- Subpart Y—Commercial diving operations
- Subpart Z—Toxic and hazardous substances

Subpart C—General Safety and Health Provisions

Table 5.1. Subpart C—General safety and health provisions training standards

Section	Designation
1926.20	General safety and health provisions
1926.21	Safety training and education
1926.35	Employee emergency action plans

Employers in Subpart C should only pay attention to the three training requirements that might be applicable. They should pay close attention to §1926.20 because it is one of the most cited construction training standards. The general provisions included in §1926.20 require construction employers to

- initiate and maintain such programs as may be necessary to comply with the Part 1926 standards and to include as a requirement of those programs a provision requiring that frequent and regular inspections be made by "competent persons" of the job sites, materials, and equipment; and
- identify (as unsafe) and prohibit the use of machinery, tools, materials, and equipment that are not in compliance with any Part 1926 standard, and limit the operation of equipment and machinery to only those employees who are qualified by training or experience to operate it.

OSHA training standard §1926.21 requires that employees be given instruction in the recognition and avoidance of unsafe conditions and the regulations applicable (to his or her work) regarding hazard control, as well as the necessary precautions and the necessary protective equipment for those employees who are required to enter "confined or enclosed spaces."

Although §1926.21 is somewhat ambiguous, it is a favorite of OSHA inspectors. Innumerable citations have been issued under this standard.

DO EMPLOYEES OPERATE EQUIPMENT OR MACHINES IN THE WORKPLACE?

Then General safety and health provisions §1926.20 training applies:

Who

This training requirement applies to all employees who operate equipment and machinery.

When

Employees must be trained prior to initial exposure and when new equipment or machinery is introduced into the workplace. The employer shall permit only those employees qualified (one who, by possession of a recognized degree, certificate, or professional standing, or who by extensive knowledge, training, and experience, has successfully demonstrated his ability to solve or resolve problems relating to the subject matter, the work, or the project) by training or experience to operate equipment and machinery. The training program shall provide frequent and regular inspections of the job sites, materials, and equipment to be made by competent persons (capable of identifying existing and predictable hazards in the surroundings or working conditions that are unsanitary, hazardous, or dangerous to employees, and who have authorization to take prompt corrective measures to eliminate them) designated by the employers.

IS SAFETY TRAINING AND EDUCATION PROVIDED TO YOUR EMPLOYEES?

Then §1926.21 training applies:

Who

This training applies to all employees in the recognition, avoidance, and prevention of unsafe conditions in the workplace. The employer shall instruct each employee in the recognition and avoidance of unsafe conditions and the regulations applicable to his work environment to control or eliminate any hazards or other exposure to illness or injury.

When

Employees who handle or use poisons, caustics, and other harmful substances shall be instructed regarding their safe handling and use, and be made aware of the potential hazards, personal hygiene, and personal protective measures required.

On job-site areas where harmful plants or animals are present, employees who might be exposed shall be instructed regarding the potential hazards and how to avoid injury, and the first aid procedures to be used in the event of injury.

Employees who handle or use flammable liquids, gases, or toxic materials shall also be instructed in the safe handling and use of these materials.

All employees who enter into confined or enclosed spaces shall be instructed in the nature of the hazards involved, the necessary precautions to be taken, and in the use of protective and emergency equipment required. "Confined or enclosed space" means any space having a limited means of egress, which is subject to the accumulation of toxic or flammable contaminants or has an oxygen-deficient atmosphere. Confined or enclosed spaces include, but are not limited to, storage tanks, process vessels, bins, boilers, ventilation or exhaust ducts, sewers, underground utility vaults, tunnels, pipelines, and open top spaces more than four feet in depth such as pits, tubs, vaults, and vessels.

DO EMPLOYEES HAVE ACCESS TO AN EMERGENCY ACTION PLAN TO ENSURE SAFETY FROM FIRES AND OTHER EMERGENCIES?

Then §1926.35 training applies:

Who

This training applies to all employees before implementing the emergency action plan; the employer shall designate a first aid provider and train a sufficient number of persons to assist in the safe and orderly emergency evacuation of employees. The employer shall review the plan with each employee covered by the plan at the following times:

- Initially when the plan is developed,
- Whenever the employee's responsibilities or designated actions under the plan change, and
- Whenever the plan is changed.

When

The employer shall review with each employee upon initial assignment those parts of the plan that the employee must know to protect the employee in the event of an emergency. The written plan shall be kept at the workplace and made available for employee review. For those employers with ten or fewer employees, the plan may be communicated orally to employees, and the employer need not maintain a written plan.

Subpart D—Occupational Health and Environmental Controls

Table 5.2.   Subpart D—Occupational health and environmental controls training standards

Section	Designation
1926.50	Medical services and first aid
1926.52	Occupational noise exposure
1926.53	Ionizing radiation
1926.54	Nonionizing radiation
1926.55	Gases, vapors, fumes, dusts, and mists
1926.57	Ventilation
1926.59	Hazard communication
1926.60	Methylenedianiline (MDA)
1926.62	Lead
1926.64	Process safety management of highly hazardous chemicals
1926.65	Hazardous waste operations and emergency response

Subpart D contains 11 training standards. The medical services and first aid requirements in §1926.50 apply at all construction sites.

Section 1926.52 is the noise standard for the construction industry. It limits employee noise exposure to 90 dBA averaged over an eight-hour day, but if you have employees exposed to high noise levels, you should also observe §1926.102 (part of Subpart E), which requires ear protective devices.

The OSHA standard §1926.53 standard applies to activities that involve ionizing radiation, radioactive materials, or X-rays. The §1926.54 standard applies when laser equipment (nonionizing radiation) is used. The use of laser safety goggles is required by a Subpart E standard, §1926.102.

If ventilation is used as an engineering control to limit exposure to airborne contaminants, then §1926.57 Ventilation standard applies.

Process safety management of highly hazardous chemicals (§1926.64) applies to unexpected releases of toxic, reactive, or flammable liquids and gases in processes involving highly hazardous chemicals.

Hazardous waste operations and emergency response (§1926.65) applies to those operations where employees are engaged in the cleanup of uncontrolled hazardous waste sites; operations involving hazardous waste treatment, storage, and disposal (TSD) facilities; emergency response operations for releases or substantial threats of releases of hazardous substances; and postemergency response operations.

The Hazard communication standard is the single most cited OSHA standard. Every employer must pay strict attention to §1926.59. It requires that chemical manufacturers and importers assess the hazards of all chemicals that they produce or import and furnish detailed material safety data sheets (MSDSs) to their customers. All employers must provide that information to their employees by means of a written hazard communication program, labels on containers, MSDSs, and employee training. The term "hazardous chemical" is defined very broadly, so virtually every employer is required to observe §1926.59.

DO EMPLOYEES HAVE ACCESS TO A MEDICAL CLINIC IN NEAR PROXIMITY TO THE PLANT OR FACILITY FOR TREATMENT?

Then Medical services and first aid §1926.50 training applies:

Who

In the absence of an infirmary, clinic, hospital, or physician, which is reasonably accessible in terms of time and distance to the worksite, and which is available for the treatment of injured employees, a person who has a valid certificate in first aid treatment of injured employees; a valid certificate in first aid training from the U.S. Bureau of Mines, the American Red Cross; or equivalent training that can be verified by documentary evidence shall be available at the worksite to render first aid.

When

Employees must be trained prior to responding to first aid or emergency situations.

ARE EMPLOYEES EXPOSED TO LOUD NOISES WHILE ON THE JOB SITE?

Then OSHA standard §1926.52 training applies:

Who

This training applies to all employees who are exposed to an eight-hour time-weighted average of 85 decibels. Protection against the effects of noise exposure should be provided when the sound levels exceed those shown when measured on the A-scale of a standard sound level meter at slow response.

When

Training should be before employees are exposed to high levels of noise and be repeated annually. When employees are subjected to sound levels exceeding those listed in table d-2 of this standard, feasible administrative or engineering controls shall be used. If controls fail to reduce sound levels within the levels of the table, personal protective equipment shall be provided and used to reduce sound levels within the levels of the table.

ARE RADIOACTIVE MATERIALS OR X-RAY EQUIPMENT USED AT YOUR FACILITY?

Then Ionizing radiation §1926.53 training applies:

Who

Any activity that involves the use of radioactive materials or X-rays, whether or not under license from the Atomic Energy Commission (Nuclear Regulatory Commission) shall be performed by competent persons specially trained in the proper and safe operation of equipment.

When

In the case of materials used under commission license, only persons actually licensed, or competent persons under the direction and supervision of the licensee, shall perform the work.

IS LASER EQUIPMENT USED AT YOUR FACILITY?

Then Nonionizing radiation §1926.54 training applies:

Who

Only qualified and trained employees shall be assigned to install, adjust, and operate laser equipment.

When

Proof of qualification of the laser equipment operator shall be available and in the possession of the operator at all times.

ARE EMPLOYEES EXPOSED TO GASES, VAPORS, FUMES, DUSTS, OR MISTS DURING CONSTRUCTION WORK?

Then §1926.55 training applies:

Who

This training applies to all employees who are exposed to air contamination that is introduced into the air having the effect of rendering the air toxic or harmful.

When

Administrative or engineering controls must first be implemented whenever feasible. When controls are not feasible to achieve full compliance, protective equipment or other protective measures shall be used to prevent the exposure of employees to air contaminants. Any equipment and technical measures used for this purpose must first be approved for each particular use by a competent industrial hygienist or other technically qualified person. Whenever respirators are used, their use shall comply with the OSHA regulation on respiratory protection.

DO EMPLOYEES WORK IN OR AROUND OPEN-SURFACE TANK OPERATIONS?

Then Ventilation §1926.57 training applies:

Who

All employees working in and around open-surface tank operations must be instructed as to the hazards of their respective jobs and in the personal protection and first aid procedures applicable to these hazards.

All persons required to work in a manner that their feet may become wet shall be provided with rubber or other impervious boots or shoes, rubbers, or wooden-soled shoes sufficient to keep feet dry.

All persons required to handle work wet with a liquid other than water shall be provided with gloves impervious to such a liquid and of a length sufficient to prevent entrance of liquid into the tops of the gloves. The interior of gloves shall be kept free from corrosive or irritating contaminants.

All persons required to work in a manner that their clothing might become wet shall be provided with such aprons, coats, jackets, sleeves, or other garments made of rubber, or of other materials impervious to liquids other than water, as are required to keep their clothing dry. Aprons shall extend well below the top of boots to prevent liquid splashing into the boots. Provision of dry, clean, cotton clothing along with rubber shoes or short boots and an apron impervious to liquids other than water shall be considered a satisfactory substitute where small parts are cleaned, plated, or acid dipped in open tanks and rapid work is required.

Whenever there is a danger of splashing, for example, when additions are made manually to the tanks, or when acids and chemicals are removed from the tanks, the employees so engaged shall be required to wear either tight-fitting chemical goggles or an effective face shield. See OSHA standard §1926.102 Eye and face protection.

When during emergencies employees must be in areas where concentrations of air contaminants are greater than the limits or oxygen concentrations are less than 19.5 percent, they must use respirators that reduce their exposure to a level below these limits or that provide adequate oxygen. The respirators must also be provided in marked, quickly accessible storage compartments built for this purpose when the possibility exists of accidental release of hazardous concentrations of air contaminants. Respirators must be approved by the National Institute for Occupational Safety and Health (NIOSH) under 42 CFR 84, selected by a competent industrial hygienist or other technically qualified source, and used in accordance with OSHA standard 29 CFR §1926.103 Respiratory protection.

When

No specific training time is mentioned in the OSHA standard.

ARE EMPLOYEES EXPOSED TO HAZARDOUS CHEMICALS WHILE ON THE JOB SITE?

Then Hazard communication §1929.59 training applies:

Who

This training applies to all employees who have exposure to hazardous chemicals while on the job site. The purpose of the Hazard communication standard is to ensure that the hazards of all chemicals produced or imported are evaluated and that information concerning their hazards is transmitted to employers and employees. This transmittal of information is to be accomplished by means of comprehensive hazard communication programs, which are to include container labeling and other forms of warning, material safety data sheets, and employee training.

When

Employers shall provide employees with effective information and training on hazardous chemicals in their work area at the time of their initial assignment and whenever a new physical or health hazard the employees have not previously been trained about is introduced into their work area. Information and training may be designed to cover categories of hazards (e.g., flammability, carcinogenicity) or specific chemicals. Chemical-specific information must always be available through labels and material safety data sheets.

Employee training should include

- methods and observations that may be used to detect the presence or release of a hazardous chemical in the work area (such as monitoring conducted by the employer, continuous monitoring devices, appearance or odor of hazardous chemicals when being released, etc.);
- the physical and health hazards of the chemicals in the work area;
- the measures employees can take to protect themselves from these hazards, including specific procedures the employer has implemented to protect employees from exposure to hazardous chemicals, such as appropriate work practices, emergency procedures, and personal protective equipment to be used; and
- the details of the hazard communication program to be developed by the employer, including an explanation of the labeling system and the material safety data sheet, and how employees can obtain and use the appropriate hazard information.

DO EMPLOYEES PERFORM CONSTRUCTION WORK WHERE METHYLENEDIANILINE (MDA) MIGHT BE EXPOSED?

Then §1926.60 training applies:

Who

This training applies to all employees when construction work in which exposure to Methylenedianiline is included in the following: construction, alteration, repair, maintenance, renovation, installation or the finishing of surfaces, MDA spill or emergency cleanup. The employer shall also provide employees with information and training on MDA in accordance with the Hazard communication standard 29 CFR §1910.1200(h) at the time of initial assignment and at least annually thereafter.

When

The employer shall provide an explanation of the contents of this standard, including appendixes A and B; indicate to employees where a copy of the standard is available; provide access to training materials; make available to all affected employees, without cost, all written materials relating to the employee training program, including a copy of this regulation; describe the medical surveillance program required; explain the information contained in appendix C; and describe the medical removal provision required under paragraph (n) of this standard.

ARE EMPLOYEES EXPOSED TO LEAD ON THE JOB SITE?

Then §1926.62 training applies:

Who

This training applies to employees who are exposed to lead hazards according to the requirements of OSHA's Hazard communication standard for the construction industry, 29 CFR §1926.59, including the requirements concerning warning signs and labels, material safety data sheets (MSDSs), and employee information and training. In addition, employers shall comply with the following requirements:

> For all employees who are subject to exposure to lead at or above the action level on any day or who are subject to exposure to lead compounds that might cause skin or eye irritation (e.g., lead arsenate, lead azide), the employer shall provide a training program and ensure employee participation.

When

The employer shall provide initial training prior to the time of job assignment or prior to the start-up date for this requirement, whichever comes last. The employer shall also provide the training program at least annually for each employee who is subject to lead exposure at or above the action level on any day.

The employer shall ensure that each employee is trained in the following:

- The content of this standard and its appendixes;
- The specific nature of the operations that could result in exposure to lead above the action level;
- The purpose, proper selection, fitting, use, and limitations of respirators;
- The purpose and a description of the medical surveillance program, and the medical removal protection program including information concerning the adverse health effects associated with excessive exposure to lead (with particular attention to the adverse reproductive effects on both males and females and hazards to the fetus and additional precautions for employees who are pregnant);
- The engineering controls and work practices associated with the employee's job assignment including training of employees to follow relevant good work practices described in appendix B in this section;
- The contents of any compliance plan in effect;
- Instructions to employees that chelating agents should not routinely be used to remove lead from their bodies and should not be used at all except under the direction of a licensed physician; and
- The employee's right of access to records under OSHA standard 29 CFR §1910.1020 Access to employee exposure and medical records.

ARE EMPLOYEES INVOLVED IN UNEXPECTED RELEASES OF TOXIC, REACTIVE, OR FLAMMABLE LIQUIDS AND GASES INVOLVING HIGHLY HAZARDOUS CHEMICALS?

Then Process safety management of highly hazardous chemicals §1926.64 training applies:

Who

Each employee presently involved in operating a process, and each employee before being involved in operating a newly assigned process, shall be trained in an overview of the process and in the operating procedures. The training shall include emphasis on the specific safety and health hazards, emergency operations including shutdown, and safe work practices applicable to the employee's job tasks.

Contract Employer Responsibilities Training

The contract employer shall ensure that each contract employee is trained in the work practices necessary to safely perform his or her job. The contract employer shall ensure that each contract employee is instructed in the known potential fire, explosion, or toxic-release hazards related to his or her job and the process, and the applicable provisions of the emergency action plan.

Mechanical Integrity Training

When training for process maintenance activities, the employer shall train each employee involved in maintaining the ongoing integrity of process equipment in an overview of that process and its hazards and in the procedures applicable to the employee's job tasks to ensure that the employee can perform the job tasks in a safe manner.

Management of Change Training

Who

Employees involved in maintenance and operating a process, and contract employees whose job tasks will be affected by a change in the process, shall be informed of, and trained in, the change prior to start-up of the process or affected part of the process.

When

In lieu of initial training for those employees already involved in operating a process, an employer may certify in writing that the employee has the required knowledge, skills, and abilities to safely carry out the duties and responsibilities as specified in the operating procedures. Refresher training shall be provided at least every three years, and more often if necessary, to each employee involved in operating a process to ensure that the employee understands and adheres to the current operating procedures of the process. The employer, in consultation with the employees involved in operating the process, shall determine the appropriate frequency of refresher training.

Training Documentation

The employer shall ascertain that each employee involved in operating a process has received and understood the training required. The employer shall prepare a record that contains the identity of the employee, the date of training, and the means used to verify that the employee understood the training.

The contract employer shall document that each contract employee has received and understood the training required by this paragraph. The contract employer shall prepare a record that contains the identity of the contract employee, the date of training, and the means used to verify that the employee understood the training.

ARE EMPLOYEES INVOLVED IN HAZARDOUS WASTE AND EMERGENCY RESPONSE OPERATIONS?

Then §1926.65 training applies:

Who

All employees who will be working on a hazardous waste site shall be informed of any risks that have been identified, and in situations covered by the Hazard communication standard, training required by that standard need not be duplicated.

All employees working on a hazardous waste site (such as but not limited to equipment operators, general laborers, and others) exposed to hazardous substances, health hazards, or safety hazards and their supervisors and management responsible for the site shall receive training meeting the requirements before they are permitted to engage in hazardous waste operations that could expose them to hazardous substances, safety, or health hazards.

Employees shall not be permitted to participate in or supervise field activities until they have been trained to a level required by their job function and responsibility. The training shall cover the following:

- Names of personnel and alternates responsible for site safety and health;
- Safety, health, and other hazards present on the site;
- Use of personal protective equipment;
- Work practices by which the employee can minimize risks from hazards;
- Safe use of engineering controls and equipment on the site;
- Medical surveillance requirements, including recognition of symptoms and signs that might indicate overexposure to hazards; and
- The contents of the site safety and health plan.

General site workers (such as equipment operators, general laborers, and supervisory personnel) engaged in hazardous substance removal or other activities that expose or potentially expose workers to hazardous substances and health hazards shall receive a minimum of 40 hours of instruction off the site and a minimum of three days actual field experience under the direct supervision of a trained, experienced supervisor.

When

Workers on site only occasionally for a specific limited task (such as, but not limited to, groundwater monitoring, land surveying, or geophysical surveying) and who are unlikely to be exposed over permissible exposure limits and published exposure limits shall receive a minimum of 24 hours of instruction off the site and a minimum of one day actual field experience under the direct supervision of a trained, experienced supervisor.

Workers regularly on site who work in areas that have been monitored and fully characterized indicating that exposures are under permissible exposure limits and published exposure limits where respirators are not necessary, and where the characterization indicates that there are no health hazards or the possibility of an emergency developing, shall receive a minimum of 24 hours of instruction off the site and a minimum of one day actual field experience under the direct supervision of a trained, experienced supervisor.

Workers with 24 hours of training and who become general site workers, or who are required to wear respirators, shall have the additional 16 hours and two days of training necessary to total the training required.

Management and Supervisor Training

Onsite management and supervisors directly responsible for, or who supervise employees engaged in, hazardous waste operations shall receive 40 hours initial training and three days of supervised field experience (the training may be reduced to 24 hours and one day if the only area of their responsibility is employees covered) and at least eight additional hours of specialized training at the time of job assignment on such topics as, but not limited to, the employer's safety and health program and the associated employee training program, personal protective equipment program, spill containment program, and health hazard monitoring procedure and techniques.

Qualifications for Trainers

Trainers shall be qualified to instruct employees about the subject matter that is being presented in training. Such trainers shall have satisfactorily completed a training program for teaching the subjects they are expected to teach, or they shall have the academic credentials and instructional experience necessary for teaching the subjects. Instructors shall demonstrate competent instructional skills and knowledge of the applicable subject matter.

Training Certification

Employees and supervisors that have received and successfully completed the training and field experience shall be certified by their instructor or the head instructor and trained supervisor as having successfully completed the necessary training. A written certificate shall be given to each person so certified. Any person who has not been certified shall be prohibited from engaging in hazardous waste operations.

Emergency Response Training

Employees who are engaged in responding to hazardous emergency situations at hazardous waste cleanup sites that might expose them to hazardous substances shall be trained in how to respond to such expected emergencies.

Refresher Training

Employees, managers, and supervisors shall receive eight hours of refresher training annually, any critique of incidents that have occurred in the past year that can serve as training examples of related work, and other relevant topics.

Equivalent Training

Employers who can show by documentation or certification that an employee's work experience or training has resulted in training equivalent to that training required shall not be required to provide the initial training requirements of those paragraphs to such employees and shall provide a copy of the certification or documentation to the employee upon request. However, certified employees or employees with equivalent training new to a site shall receive appropriate, site-specific training before site entry and have appropriate supervised field experience at the new site. Equivalent training includes any academic training or the training that existing employees might have already received from actual hazardous waste site work experience.

Subpart E—Personal Protective and Life Saving Equipment

Table 5.3. Subpart E—Personal protective and life saving equipment training standards

Section	Designation
1926.101	Hearing protection
1926.103	Respiratory protection

There are only two Subpart E training standards to be concerned with. They state in very general terms the conditions under which protective equipment must be used.

ARE EMPLOYEES EXPOSED TO HIGH LEVELS OF NOISE?

Then Hearing protection §1926.101 training applies:

Who

Wherever the noise levels or duration of exposures are too high, ear protective devices shall be provided and used.

When

Ear protective devices inserted in the ear shall be fitted or determined individually by competent persons.

DO EMPLOYEES USE RESPIRATORS?

Then Respiratory protection §1926.103 training applies:

Who

In any workplace where respirators are necessary to protect the health of the employee or whenever respirators are required by the employer, the employer shall establish and implement a written respiratory protection program with worksite-specific procedures. The program shall be updated as necessary to reflect those changes in workplace conditions that affect respirator use. The employer shall include in the program the following:

- Procedures for selecting respirators for use in the workplace;
- Medical evaluations of employees required to use respirators;
- Fit testing procedures for tight-fitting respirators;
- Procedures for proper use of respirators in routine and reasonably foreseeable emergency situations;
- Procedures and schedules for cleaning, disinfecting, storing, inspecting, repairing, discarding, and otherwise maintaining respirators;
- Procedures to ensure adequate air quality, air quantity, and flow of breathing air for atmosphere-supplying respirators;
- Training of employees in the respiratory hazards to which they are potentially exposed during routine and emergency;
- Training of employees in the proper use of respirators, including putting on and removing them, any limitations on their use, and their maintenance; and
- Procedures for regularly evaluating the effectiveness of the program.

The employer shall provide respirators, training, and medical evaluations at no cost to the employee.

For all immediately dangerous to life or health (IDLH) atmospheres, the employer shall ensure that the employees located outside the IDLH atmosphere are trained and equipped to provide effective emergency rescue.

Training and Information

The employer must provide effective training to employees who are required to use respirators. The training must be comprehensive, be understandable, and recur annually and more often if necessary. The employer shall ensure that each employee can demonstrate knowledge of at least the following:

- Why the respirator is necessary and how improper fit, usage, or maintenance can compromise the protective effect of the respirator;
- What the limitations and capabilities of the respirator are;
- How to use the respirator effectively in emergency situations, including situations in which the respirator malfunctions;
- How to inspect, put on and remove, use, and check the seals of the respirator;
- What the procedures are for maintenance and storage of the respirator; and
- How to recognize medical signs and symptoms that may limit or prevent the effective use of respirators

The training shall be conducted in a manner that is understandable to the employee.

When

The employer shall provide the training prior to requiring the employee to use a respirator in the workplace. An employer who is able to demonstrate that a new employee has received training within the last 12 months is not required to repeat this training provided that the employee can demonstrate knowledge of those elements. Previous training not repeated initially by the employer must be provided no later than 12 months from the date of the previous training.

Retraining shall be administered annually and when the following situations occur:

- Changes in the workplace or the type of respirator render previous training obsolete;
- Inadequacies in the employee's knowledge or use of the respirator indicate that the employee has not retained the requisite understanding or skill; or
- Any other situation arises in which retraining appears necessary to ensure safe respirator use.

Subpart F—Fire Protection and Prevention

Table 5.4. Subpart F—Fire protection and prevention training standards

Section	Designation
1926.150	Fire protection
1926.152	Flammable and combustible liquids

There are two Subpart F training standards. The fire protection requirements in §1926.150 are rather detailed and should be read and observed by all construction employers.

Among other things, those standards require the development of a fire protection program (§1926.150[a]), a trained fire brigade "as warranted by the project" (§1926.150[a][5]), and periodic inspection and maintenance of all portable fire extinguishers (§1926.150[c][1][viii]).

If flammable or combustible liquids are present at a construction site, the detailed requirements of §1926.152 must be observed.

IS A FIRE PROTECTION PROGRAM PROVIDED THROUGHOUT ALL PHASES OF CONSTRUCTION WORK?

Then §1926.150 training applies:

Who

As needed for a project, the employer shall provide a trained and equipped fire-fighting organization (fire brigade) to assure adequate protection to life. "Fire brigade" means an organized group of employees that are knowledgeable, trained, and skilled in the safe evacuation of employees during emergency situations and in assisting in fire-fighting operations.

When

Portable fire extinguishers shall be inspected periodically and maintained in accordance with Maintenance and use of portable fire extinguishers.

The owner or occupant of a property in which fire extinguishers are located has an obligation for the care and use of these extinguishers at all times. By doing so, he is contributing to the protection of life and property. The nameplates and instruction manual should be read and thoroughly understood by all persons who might be expected to use extinguishers.

To discharge this obligation he should give proper attention to the inspection, maintenance, and recharging of this fire protective equipment. He should also train his personnel in the correct use of fire extinguishers on the different types of fires that might occur on his property.

DOES YOUR COMPANY HANDLE OR STORE FLAMMABLE OR COMBUSTIBLE LIQUIDS?

Then §1926.152 training applies:

Who

This training applies to all employees who are exposed to flammable and combustible liquids. Detailed, printed instructions of what to do in flood emergencies are properly posted. Station operators and other employees depended upon to carry out such instructions are thoroughly informed as to the location and operation of such valves and other equipment necessary to effect these requirements.

When

No specific training time is mentioned in the OSHA standard.

Subpart I—Tools (Hand and Power)

Table 5.5.   Subpart I—Tools (hand and power) training standards

Section	Designation
1926.302	Power-operated hand tools
1926.304	Woodworking tools

There are two training requirements in Subpart I. They both set forth the requirements to be observed whenever the different kinds of tools and equipment are used in construction.

DO EMPLOYEES USE ANY HAND OR POWER TOOLS?

Then §1926.302 training applies:

Who

All employees who have been trained in the operation of the particular tool in use shall be allowed to operate a powder-actuated tool.

The tool shall be tested each day before loading to see that safety devices are in proper working condition. The method of testing shall be in accordance with the manufacturer's recommended procedure.

Any tool found not in proper working order, or that develops a defect during use, shall be immediately removed from service and not used until properly repaired.

Personal protective equipment shall be required.

Tools shall not be loaded until just prior to the intended firing time. Neither loaded nor empty tools are to be pointed at any employees. Hands shall be kept clear of the open barrel end.

When

Employees must receive training before they are allowed to operate hand and power tools.

DO EMPLOYEES USE WOODWORKING TOOLS?

Then §1926.304 training applies:

Who

This training applies to all workers permitted to operate woodworking tools.

When

Before an employee is permitted to operate any woodworking machine, he shall receive instructions in the hazards of the machine and the safe method of its operation. Training of operators of machines, tools, and equipment shall apply as follows:

- Learn the machine's applications and limitations, as well as the specific potential hazards peculiar to this machine. Follow available operating instructions and safety rules carefully.
- Keep working area clean and be sure adequate lighting is available.
- Do not wear loose clothing, gloves, bracelets, necklaces, or ornaments. Wear face, eye, ear, respiratory, and body protection devices, as indicated for the operation or environment.
- Do not use cutting tools larger or heavier than the machine is designed to accommodate. Never operate a cutting tool at greater speed than recommended.
- Keep hands well away from saw blades and other cutting tools. Use a push stock or push block to hold or guide the work when working close to a cutting tool.
- Whenever possible use properly locked clamps, a jig, or a vise to hold the work.
- Combs (feather boards) shall be provided for use when an applicable guard cannot be used.
- Never stand directly in line with a horizontally rotating cutting tool. This is particularly true when first starting a new tool or a new tool is initially installed on the arbor.
- Be sure the power is disconnected from the machine before tools are serviced.
- Never leave the machine with the power on.
- Be positive that hold-downs and antikickback devices are positioned properly and that the work piece is being fed through the cutting tool in the right direction.
- Do not use a dull, gummy, bent, or cracked cutting tool.
- Be sure that keys and adjusting wrenches have been removed before turning power on.
- And use only accessories designed for the machine and adjust the machine for minimum exposure of the cutting tool necessary to perform the operation.

Subpart J—Welding and Cutting

Table 5.6.   Subpart J—Welding and cutting training standards

Section	Designation
1926.350	Gas welding and cutting
1926.351	Arc welding and cutting
1926.352	Fire prevention

Subpart J has three training requirements. They regulate all aspects of welding, cutting, and heating when those operations are performed on a construction project. Employers should pay careful attention to Subpart J standards whenever welding or cutting is done.

IS GAS WELDING OR CUTTING PERFORMED BY YOUR EMPLOYEES?

Then §1926.350 training applies:

Who

This training applies to all employees who have duties involving gas welding and cutting.

When

No specific training time is mentioned in the OSHA standard.

The employer shall thoroughly instruct employees in the safe use of fuel gas as follows: before a regulator to a cylinder valve is connected, the valve shall be opened slightly and closed immediately. (This action is generally termed "cracking" and is intended to clear the valve of dust or dirt that might otherwise enter the regulator.) The person cracking the valve shall stand to one side of the outlet, not in front of it.

The valve of a fuel gas cylinder shall not be cracked where the gas would reach welding work, sparks, flame, or other possible sources of ignition. The cylinder valve shall always be opened slowly to prevent damage to the regulator.

For quick closing, valves on fuel gas cylinders shall not be opened more than one and a half turns. When a special wrench is required, it shall be left in position on the stem of the valve while the cylinder is in use so that the fuel gas flow can be shut off quickly in case of an emergency. In the case of manifold or coupled cylinders, at least one such wrench shall always be available for immediate use. Nothing shall be placed on top of a fuel gas cylinder, when in use, which may damage the safety device or interfere with the quick closing of the valve.

Fuel gas shall not be used from cylinders through torches or other devices that are equipped with shutoff valves without reducing the pressure through a suitable regulator attached to the cylinder valve or manifold. Before a regulator is removed from a cylinder valve, the cylinder valve shall always be closed and the gas released from the regulator.

If the valve on a fuel gas cylinder is opened and there is found to be a leak around the valve stem, the valve shall be closed and the gland nut tightened. If this action does not stop the leak, the use of the cylinder shall be discontinued, and it shall be properly tagged and removed from the work area.

In the event that fuel gas should leak from the cylinder valve, rather than from the valve stem, and the gas cannot be shut off, the cylinder shall be properly tagged and removed from the work area. If a regulator attached to a cylinder valve will effectively stop a leak through the valve seat, the cylinder need not be removed from the work area. If a leak should develop at a fuse plug or other safety device, the cylinder shall be removed from the work area.

IS ARC WELDING OR CUTTING PERFORMED BY YOUR EMPLOYEES?

Then §1926.351 training applies:

Who

This training applies to all employees who have duties involving arc welding and cutting.

When

No specific training time is mentioned in the OSHA standard.
   Employers shall instruct employees in the safe means of arc welding and cutting as follows:

- When electrode holders are to be left unattended, the electrodes shall be removed, and the holders shall be so placed or protected so that they cannot make electrical contact with employees or conducting objects.
- Hot electrode holders shall not be dipped in water; to do so may expose the arc welder or cutter to electric shock.
- And when the arc welder or cutter has occasion to leave his work or to stop work for any appreciable length of time, or when the arc welding or cutting machine is to be moved, the power supply switch to the equipment shall be opened.

ARE EMPLOYEES ASSIGNED TO GUARD AGAINST FIRE DURING WELDING, CUTTING, OR HEATING OPERATIONS?

Then Fire prevention §1926.352 training applies:

Who

Employees assigned to guard against fire during welding, cutting, or heating operations (and for a sufficient period of time after completion of the work) are to be instructed on the specific anticipated fire hazards and how the provided fire-fighting equipment is to be used.

When

No training time is mentioned in the OSHA standard.

Subpart K—Electrical

Table 5.7. Subpart K—Electrical training standards

Installation safety requirements	
Section	Designation
1926.404	Wiring design and protection

Subpart K covers the electrical safety requirements for construction job sites. They are separated into four major divisions and are very complicated. However, there is only one training requirement to be concerned with.

DO EMPLOYEES PERFORM ANY ELECTRICAL WIRING DESIGN AT THE JOB SITE?

Then §1926.404 training applies:

Who

This training applies to workers who are exposed to electrical hazards on the job. The employer shall use either ground-fault circuit interrupters or assured equipment-grounding conductor program to protect employees on construction sites. The employer shall designate one or more competent persons to implement the program.

When

No training time is mentioned in the OSHA standard.

Subpart L—Scaffolding

Table 5.8. Subpart L—Scaffolding training standards

Section	Designation
1926.453	Aerial lifts
1926.454	Scaffolds training requirements

Subpart L Scaffolding has two training requirements.

OSHA standard §1926.453 sets some specific requirements for aerial lift operations. It applies to equipment for vehicle-mounted elevating and rotating work platforms.

Employers who use scaffolds in construction work should be familiar with all the training requirements included in §1926.454. It contains all the general requirements for scaffolds. It also sets the minimum strength criteria for all scaffold components and connections. It requires that each scaffold component be capable of supporting, without failure, its own weight and at least four times the maximum intended load applied to it.

OSHA standard §1926.454 distinguishes between the training needed by employees to erect and to dismantle scaffolds.

The standard applies to all construction work. It requires employers to instruct each employee in the recognition and avoidance of unsafe conditions. It also sets certain criteria allowing employers to tailor training to fit the particular circumstances of each employer's workplace.

DO EMPLOYEES USE AERIAL LIFTS?

Then §1926.453 training applies:

Who

Lift controls shall be tested each day prior to use to determine that such controls are in safe working condition. Only authorized persons shall operate an aerial lift.

When

Before moving an aerial lift for travel, the booms shall be inspected to see that it is properly cradled and outriggers are in stowed position.

DO EMPLOYEES USE SCAFFOLDS?

Then §1926.454 training applies:

Who

This training applies to all employees who perform work while on a scaffold. These employees should be trained by a person qualified in the subject matter to recognize the hazards associated with the type of scaffold being used and to understand the procedures to control or minimize those hazards. The training shall include the following areas:

- The nature of any electrical hazards, fall hazards, and falling object hazards in the work area;
- The correct procedures for dealing with electrical hazards and for erecting, maintaining, and disassembling the fall protection systems and falling object protection systems being used;
- The proper use of the scaffold and the proper handling of materials on the scaffold; and
- The maximum intended load and the load-carrying capacities of the scaffolds used.

The employer shall have each employee who is involved in erecting, disassembling, moving, operating, repairing, maintaining, or inspecting a scaffold trained by a competent person to recognize any hazards associated with the work in question. The training shall include the following topics:

- The nature of scaffold hazards;
- The correct procedures for erecting, disassembling, moving, operating, repairing, inspecting, and maintaining the type of scaffold in question; and
- The design criteria, maximum intended load-carrying capacity, and intended use of the scaffold.

When

When the employer has reason to believe that an employee lacks the skill or understanding needed for safe work involving the erection, use, or dismantling of scaffolds, the employer shall retrain each such employee so that the requisite proficiency is regained. Retraining is required in the following situations:

- Where changes at the worksite present a hazard about which an employee has not been previously trained;
- Where changes in the types of scaffolds, fall protection, falling object protection, or other equipment present a hazard about which an employee has not been previously trained; or
- Where inadequacies in an affected employee's work involving scaffolds indicate that the employee has not retained the requisite proficiency.

Subpart M—Fall Protection

Table 5.9. Subpart M—Fall protection training standards

Section	Designation
1926.503	Fall protection training requirements

Subpart M has only one training requirement to be concerned with. The Fall protection standard contains OSHA's regulations for construction workplaces to prevent employees from falling off, onto, or through levels, and to protect employees from being struck by falling objects.

Under the standard, employers are able to select fall protection measures compatible with the type of work being performed. Fall protection generally can be provided through the use of guardrail systems, safety net systems, personal fall arrest systems, positioning device systems, and warning line systems.

OSHA standard §1926.503 covers Fall protection training requirements. Employers must provide a training program that teaches employees who might be exposed to fall hazards how to recognize such hazards and how to minimize them.

DO EMPLOYEES USE FALL PROTECTION SYSTEMS DURING CONSTRUCTION WORK?

Then §1926.503 training applies:

Who

The employer shall provide a training program for each employee who might be exposed to fall hazards. The program will enable each employee to recognize the hazards of falling and train each employee in the procedures to be followed in order to minimize these hazards.

The employer will ensure that each employee has been trained, as necessary, by a competent person qualified in the following areas:

- The nature of fall hazards in the work area;
- The correct procedures for erecting, maintaining, disassembling, and inspecting the fall protection systems to be used;
- The use and operation of guardrail systems, personal fall arrest systems, safety net systems, warning line systems, safety monitoring systems, controlled access zones, and other protection to be used;
- The role of each employee in the safety monitoring system when this system is used;
- The limitations on the use of mechanical equipment during the performance of roofing work on low-sloped roofs; and
- The correct procedures for the handling and storage of equipment and materials and the erection of overhead protection.

When

The employer shall verify compliance by preparing a written certification record. The written certification record shall contain the name or other identity of the employee trained, the date of the training, and the signature of the person who conducted the training or the signature of the employer.

If the employer relies on training conducted by another employer or completed prior to the effective date of this section, the certification record shall indicate the date the employer determined the prior training was adequate rather than the date of actual training, and the latest training certification shall be maintained.

Retraining

When the employer has reason to believe that any affected employee who has already been trained does not have the necessary understanding and skill, the employer shall retrain each employee. Circumstances where retraining is required include the following:

- Changes in the workplace render previous training obsolete.
- Changes in the types of fall protection systems or equipment to be used render previous training obsolete.
- Inadequacies in an affected employee's knowledge or use of fall protection systems or equipment indicate that the employee has not retained the requisite understanding or skill.

Subpart N—Cranes, Derricks, Hoists, Elevators, and Conveyors

Table 5.10.   Subpart N—Cranes, derricks, hoists, elevators, and conveyors training standards

Section	Designation
1926.550	Cranes and derricks
1926.552	Material hoists, personnel hoists, and elevators

Subpart N contains two training requirements. Construction employers who use lifting equipment must familiarize themselves with the applicable training requirements and observe them.

DO EMPLOYEES OPERATE CRANES OR DERRICKS?

Then §1926.550 training applies:

Who

This training applies to all employees who operate cranes and derricks. The employer shall comply with the manufacturer's specifications and limitations applicable to the operation of any and all cranes and derricks.

Where manufacturer's specifications are not available, the limitations assigned to the equipment shall be based on the determinations of a qualified engineer competent in this field, and such determinations will be appropriately documented and recorded. Attachments used with cranes shall not exceed the capacity, rating, or scope recommended by the manufacturer.

The employer shall designate a competent person who shall inspect all machinery and equipment prior to each use, and during use, to make sure it is in safe operating condition. Any deficiencies shall be repaired, or defective parts replaced, before continued use.

A thorough, annual inspection of the hoisting machinery shall be made by a competent person, or by a government or private agency recognized by the U.S. Department of Labor. The employer shall maintain a record of the dates and results of inspections for each hoisting machine and piece of equipment.

Personnel Platforms

Personnel platforms and suspension systems shall be designed by a qualified engineer or a qualified person competent in structural design.

When

A visual inspection of the crane or derrick, rigging, personnel platform, and the crane or derrick base support or ground shall be conducted by a competent person immediately after the trial lift to determine whether the testing has exposed any defect or produced any adverse effect upon any component or structure.

DO EMPLOYEES OPERATE MATERIAL HOISTS, PERSONNEL HOISTS, OR ELEVATORS?

Then §1926.552 training applies:

Who

This training applies to all employees who operate material hoists, personnel hoists, and elevators. The employer shall comply with the manufacturer's specifications and limitations applicable to the operation of all hoists and elevators. Where manufacturer's specifications are not available, the limitations assigned to the equipment shall be based on the determinations of a professional engineer competent in the field.

All material hoist towers shall be designed by a licensed professional engineer.

When

Following assembly and erection of hoists, and before being put in service, an inspection and test of all functions and safety devices shall be made under the supervision of a competent person. A similar inspection and test is required following major alteration of an existing installation. All hoists shall be inspected and tested at not more than three-month intervals. Records shall be maintained and kept on file for the duration of the job.

Personnel hoists used in bridge tower construction shall be approved by a registered professional engineer and erected under the supervision of a qualified engineer competent in this field.

Subpart O—Motor Vehicles, Mechanized Equipment, and Marine Operations

Table 5.11. Subpart O—Motor vehicles, mechanized equipment, and marine operations training standards

Section	Designation
1926.602	Material handling equipment
1926.604	Site clearing

There are two training requirements in Subpart O standards that apply to equipment and operations.

OSHA standard §1926.602 applies to earthmoving equipment, off-highway trucks, rollers, compactors, front-end loaders, bulldozers, tractors, lift trucks, stackers, high-lift rider industrial trucks, and similar construction equipment.

Section 1926.604 requires rollover guards and canopy guards on equipment used in site-clearing operations and provides that employees engaged in such operations be protected from irritant and toxic plants and instructed in the first aid treatment that is available.

DO EMPLOYEES OPERATE POWERED INDUSTRIAL TRUCKS (FORKLIFT TRUCKS)?

Then §1926.602 training applies:

Who

The employer shall ensure that each powered industrial truck operator is competent to operate a powered industrial truck safely, as demonstrated by the successful completion of the training and evaluation.

All operator training and evaluation shall be conducted by persons who have the knowledge, training, and experience to train powered industrial truck operators and evaluate their competence.

Powered industrial truck operators shall receive initial training in the following topics, except in topics that the employer can demonstrate are not applicable to safe operation of the truck in the employer's workplace.

When

Prior to permitting an employee to operate a powered industrial truck, the employer shall ensure that each operator has successfully completed the training.

Refresher Training

Refresher training, including an evaluation of the effectiveness of that training, shall be conducted to ensure that the operator has the knowledge and skills needed to operate the powered industrial truck safely. Refresher training shall be provided to the operator when

- the operator has been observed to operate the vehicle in an unsafe manner;
- the operator has been involved in an accident or near miss incident;
- the operator has received an evaluation that reveals that the operator is not operating the truck safely;
- the operator is assigned to drive a different type of truck; or
- a condition in the workplace changes in a manner that could affect safe operation of the truck.

An evaluation of each powered industrial truck operator's performance shall be conducted at least once every three years.

Avoidance of Duplicative Training

If an operator has previously received training in a topic and such training is appropriate to the truck and working conditions encountered, additional training in that topic is not required if the operator has been evaluated and found competent to operate the truck safely.

Certification of Training

The employer shall certify that each operator has been trained and evaluated. The certification shall include the name of the operator, the date of the training, the date of the evaluation, and the identity of the person performing the training.

Training shall consist of a combination of formal instruction (e.g., lecture, discussion, interactive computer learning, video tape, written material), practical training (demonstrations performed by the trainer and practical exercises performed by the trainee), and evaluation of the operator's performance in the workplace.

Truck-Related Training Topics

- Operating instructions, warnings, and precautions for the types of truck the operator will be authorized to operate
- Differences between the truck and the automobile
- Truck controls and instrumentation: where they are located, what they do, and how they work
- Engine or motor operation
- Steering and maneuvering
- Visibility (including restrictions due to loading)
- Fork and attachment adaptation, operation, and use limitations
- Vehicle capacity
- Vehicle stability
- Any vehicle inspection and maintenance that the operator will be required to perform
- Refueling and charging and recharging of batteries
- Operating limitations
- Any other operating instructions, warnings, or precautions listed in the operator's manual for the types of vehicle that the employee is being trained to operate

Workplace-Related Training Topics

- Surface conditions where the vehicle will be operated
- Composition of loads to be carried and load stability
- Load manipulation, stacking, and unstacking
- Pedestrian traffic in areas where the vehicle will be operated
- Narrow aisles and other restricted places where the vehicle will be operated
- Hazardous (classified) locations where the vehicle will be operated
- Ramps and other sloped surfaces that could affect the vehicle's stability

- Closed environments and other areas where insufficient ventilation or poor vehicle maintenance could cause a buildup of carbon monoxide or diesel exhaust
- Other unique or potentially hazardous environmental conditions in the workplace that could affect safe operation

ARE EMPLOYEES ENGAGED IN SITE-CLEARING OPERATIONS?

Then §1926.604 training applies:

Who

All employees engaged in site-clearing operations must be protected from the hazards of irritant and toxic plants.

When

Employees will be trained and instructed in the first aid treatment available prior to initial exposure.

Subpart P—Excavations

Table 5.12. Subpart P—Excavations training standards

Section	Designation
1926.651	Specific excavation requirements
1926.652	Requirements for protective systems

Subpart P standards apply to all open excavations (including trenches) made in the earth's surface. In some situations, it requires the use of written designs (approved by a registered professional engineer) for sloping, benching, and support systems. There are only two training requirements in Subpart P.

ARE YOUR EMPLOYEES INVOLVED IN EXCAVATION WORK?

Then §1926.651 training applies:

Who

This training applies to all construction employees who will be working in excavations.

When

Daily inspections of excavations, the adjacent areas, and protective systems shall be made by a competent person for evidence of a situation that could result in possible cave-ins, indications of failure of protective systems, hazardous atmospheres, or other hazardous conditions. An inspection shall be conducted by the competent person prior to the start of work and as needed throughout the shift. Inspections shall also be made after every rainstorm or other hazard-increasing occurrence. These inspections are only required when employee exposure can be anticipated.

Where the competent person finds evidence of a situation that could result in a possible cave-in, a failure of protective systems, hazardous atmospheres, or other hazardous conditions, exposed employees will be removed from the hazardous area until the necessary precautions have been taken to ensure their safety.

Structural ramps that are used solely by employees as a means of access or egress from excavations shall be designed by a competent person. Structural ramps used for access or egress of equipment shall be designed by a competent person qualified in structural design and shall be constructed in accordance with the design.

Protection from Hazards Associated with Water Accumulation

If water is controlled or prevented from accumulating by the use of water-removal equipment, the water-removal equipment and operations shall be monitored by a competent person to ensure proper operation.

If excavation work interrupts the natural drainage of surface water (such as streams), diversion ditches, dikes, or other suitable means shall be used to prevent surface water from entering the excavation and to provide adequate drainage of the area adjacent to the excavation. Excavations subject to runoff from heavy rains will require an inspection by a competent person.

Stability of Adjacent Structures

Where the stability of adjoining buildings, walls, or other structures is endangered by excavation operations, support systems such as shoring, bracing, or underpinning shall be provided to ensure the stability of such structures for the protection of employees.

A registered professional engineer must approve the determination that the structure is sufficiently removed from the excavation so as to be unaffected by the excavation activity, or a registered professional engineer must approve the determination that such excavation work will not pose a hazard to employees.

ARE EMPLOYEES PROTECTED FROM CAVE-INS BY ADEQUATE PROTECTIVE SYSTEMS?

Then §1926.652 training applies:

Who

This training applies to all construction employees who will be working in excavations.

When

Daily inspections of excavations, the adjacent areas, and protective systems shall be made by a competent person for evidence of a situation that could result in possible cave-ins or a failure of protective systems.

Materials and equipment used for protective systems shall be free from damage or defects that might impair their proper function.

Manufactured materials and equipment used for protective systems shall be used and maintained in a manner that is consistent with the recommendations of the manufacturer, and in a manner that will prevent employee exposure to hazards.

When material or equipment that is used for protective systems is damaged, a competent person shall examine the material or equipment and evaluate its suitability for continued use. If the competent person cannot assure that the material or equipment is able to support the intended loads or is otherwise suitable for safe use, then such material or equipment shall be removed from service and shall be evaluated and approved by a registered professional engineer before being returned to service.

Subpart Q—Concrete and Masonry Construction

Table 5.13. Subpart Q—Concrete and masonry construction training standards

Section	Designation
1926.701	General requirements

Subpart Q includes detailed requirements to be observed in concrete and masonry construction operations. There is only one training requirement applicable in this subpart.

DO EMPLOYEES ENGAGE IN CONCRETE AND MASONRY CONSTRUCTION WORK?

Then §1926.701 training applies:

Who

This training applies to all construction employees who will be working in concrete and masonry construction.

When

No employee, except those essential to the jacking operation, shall be permitted in the building or structure while any jacking operation is taking place unless the building or structure has been reinforced sufficiently to ensure its integrity during erection.

The phrase "reinforced sufficiently to ensure its integrity" means that a registered professional engineer, independent of the engineer who designed and planned the lifting operation, has determined from the plans that, if there is a loss of support at any jack location, that loss will be confined to that location and the structure as a whole will remain stable.

All welding on temporary and permanent connections shall be performed by a certified welder, familiar with the welding requirements specified in the plans and specifications for the lift-slab operation.

No construction loads shall be placed on a concrete structure or portion of a concrete structure unless the employer determines, based on information received from a person who is qualified in structural design, that the structure or portion of the structure is capable of supporting the loads.

The design of the shoring shall be prepared by a qualified designer, and the erected shoring shall be inspected by an engineer qualified in structural design. Lift-slab operations shall be designed and planned by a registered professional engineer who has experience in lift-slab construction. Such plans and designs shall be implemented by the employer and shall include detailed instructions and sketches indicating the prescribed method of erection. These plans and designs shall also include provisions for ensuring lateral stability of the building or structure during construction.

Subpart R—Steel Erection

Table 5.14. Subpart R—Steel erection training standards

Section	Designation
1926.753	Hoisting and rigging
1926.754	Structural steel assembly
1926.755	Column anchorage
1926.756	Beams and columns
1926.757	Open web steel joists
1926.758	Systems-engineered metal buildings
1926.759	Falling object protection
1926.761	Training

Subpart R Steel erection covers hazards associated with working under loads; hoisting, landing and placing decking; column stability; double connections; landing and placing steel joints; and falls to lower levels.

There are eight training requirements to be concerned with in this subpart.

ARE YOUR EMPLOYEES INVOLVED WITH HOISTING AND RIGGING OPERATIONS?

Then §1926.753 training applies:

Who

This training applies to all employees who are involved in hoisting and rigging operations. The employer shall also comply with the manufacturer's specifications and limitations applicable to the operation of any and all cranes and derricks.

Where manufacturer's specifications are not available, the limitations assigned to the equipment shall be based on the determinations of a qualified engineer competent in this field, and such determinations will be appropriately documented and recorded. Attachments used with cranes shall not exceed the capacity, rating, or scope recommended by the manufacturer.

When

The employer shall designate a competent person who shall inspect all machinery and equipment prior to each use, and during use, to make sure it is in safe operating condition. Any deficiencies shall be repaired, or defective parts replaced, before continued use.

A thorough annual inspection of the hoisting machinery shall be made by a competent person. The employer shall maintain a record of the dates and results of inspections for each hoisting machine and piece of equipment.

Preshift Visual Inspection of Cranes

Cranes being used in steel erection activities shall be visually inspected prior to each shift by a competent person; the inspection shall include observation for deficiencies during operation. The following items must be inspected:

- All control mechanisms for maladjustments;
- Control and drive mechanism for excessive wear of components and contamination by lubricants, water, or other foreign matter;
- Safety devices, including but not limited to boom angle indicators, boom stops, boom kick-out devices, anti-two-block devices, and load moment indicators where required;
- Air, hydraulic, and other pressurized lines for deterioration or leakage, particularly those that flex in normal operation;
- Hooks and latches for deformation, chemical damage, cracks, or wear;
- Wire rope reeving for compliance with hoisting equipment manufacturer's specifications;
- Electrical apparatus for malfunctioning, signs of excessive deterioration, dirt, or moisture accumulation;
- Hydraulic system for proper fluid level;
- Tires for proper inflation and condition;
- Ground conditions around the hoisting equipment for proper support, including ground settling under and around outriggers, groundwater accumulation, or similar conditions; and
- The hoisting equipment for level position after each move and setup.

If any deficiency is identified, an immediate determination shall be made by the competent person as to whether the deficiency constitutes a hazard.

A qualified rigger (a rigger who is also a qualified person) shall inspect the rigging prior to each shift.

Multiple Lift Rigging Training

The employer shall ensure that each employee who performs multiple lift rigging has been given training in the following areas:

- Using a multiple lift rigging assembly;
- The maximum allowed members hoisted per lift (five);
- Items lifted (only beams and similar structural members);
- The nature of the hazards associated with multiple lifts; and
- The proper procedures and equipment to perform multiple lifts.

No crane is permitted to be used for a multiple lift where such use is contrary to the manufacturer's specifications and limitations.

ARE EMPLOYEES INVOLVED IN STRUCTURAL STEEL ASSEMBLY WORK?

Then §1926.754 training applies:

Who

This training applies to all employees who are involved in structural steel assembly.

When

When deemed necessary by a competent person, plumbing-up equipment shall be installed in conjunction with the steel erection process to ensure the stability of the structure.

When used, plumbing-up equipment shall be in place and properly installed before the structure is loaded with construction material such as loads of joists, bundles of decking, or bundles of bridging.

Plumbing-up equipment shall be removed only with the approval of a competent person.

ARE EMPLOYEES INVOLVED IN COLUMN ANCHORAGE WORK?

Then §1926.755 training applies:

Who

This training applies to all employees who are involved in column anchorage.

All columns shall be evaluated by a competent person to determine whether guying or bracing is needed; if guying or bracing is needed, it shall be installed.

Anchor rods (anchor bolts) shall not be repaired, replaced, or field modified without the approval of the project structural engineer of record.

When

Prior to the erection of a column, the controlling contractor shall provide written notification to the steel erector if there has been any repair, replacement, or modification of the anchor rods (anchor bolts) of that column.

ARE EMPLOYEES INVOLVED WITH BEAMS AND COLUMNS WORK?

Then §1926.756 training applies:

Who

This training applies to all employees who work with beams and columns.

When

A competent person shall determine if more than two bolts are necessary to ensure the stability of cantilevered members; if additional bolts are needed, they shall be installed.

DO EMPLOYEES WORK WITH OPEN WEB STEEL JOISTS?

Then §1926.757 training applies:

Who

This training applies to all employees who are involved with open web steel joists.

Where steel joists at or near columns span more than 60 feet, the joists shall be set in tandem with all bridging installed unless an alternative method of erection, which provides equivalent stability to the steel joist, is designed by a qualified person and is included in the site-specific erection plan.

No modification that affects the strength of a steel joist or steel joist girder shall be made without the approval of the project structural engineer of record.

Steel joists and steel joist girders shall not be used as anchorage points for a fall arrest system unless written approval to do so is obtained from a qualified person.

When

The employer has first determined from a qualified person and documented in a site-specific erection plan that the structure or portion of the structure is capable of supporting the load.

DO EMPLOYEES WORK WITH SYSTEMS-ENGINEERED METAL BUILDINGS?

Then §1926.758 training applies:

Who

This training applies to all employees who work with systems-engineered metal buildings.

When

Purlins and girts shall not be used as an anchorage point for a fall arrest system unless written approval is obtained from a qualified person.

DO EMPLOYEES WORK IN OPERATIONS WITH FALLING OBJECT PROTECTION?

Then §1926.759 training applies:

Who

This training applies to all employees who work in steel erection operations.

When

The controlling contractor shall bar other construction processes below steel erections unless overhead protection for the employees below is provided.

ARE EMPLOYEES INVOLVED IN STEEL ERECTION OPERATIONS?

Then §1926.761 training applies:

Who

This training applies to all employees involved in steel erection operations; steel erections operations include the following: multiple lifting procedures, connector procedure, controlled decking zone procedures, hoisting and rigging, structural steel assembly, column anchorage, and working on beams and columns, open web steel joists, and systems-engineered metal buildings. Required training shall be provided by a qualified person.

When

The employer shall provide a training program for all employees exposed to fall hazards. The program shall include training and instruction in the following areas:

- The recognition and identification of fall hazards in the work area;
- The use and operation of guardrail systems (including perimeter safety cable systems), personal fall arrest systems, positioning device systems, fall restraint systems, safety net systems, and other protection to be used;
- The correct procedures for erecting, maintaining, disassembling, and inspecting the fall protection systems to be used; and
- The procedures to be followed to prevent falls to lower levels and through or into holes and openings in walking/working surfaces and walls.

Subpart S—Underground Construction, Caissons, Cofferdams, and Compressed Air

Table 5.15.   Subpart S—Underground construction, caissons, cofferdams, and compressed air training standards

Section	Designation
1926.800	Underground construction
1926.803	Compressed air

Subpart S applies only to underground construction work including tunnels, shafts, chambers, and passageways; caisson work; work on cofferdams; and work conducted in a compressed air environment. There are only two training requirements applicable in this subpart.

ARE EMPLOYEES INVOLVED IN UNDERGROUND CONSTRUCTION WORK?

Then §1926.800 training applies:

Who

This training applies to all employees involved in underground construction. All employees shall be instructed in the recognition and avoidance of hazards associated with underground construction activities including, where appropriate, the following subjects:

- Air monitoring;
- Ventilation;
- Illumination;
- Communications;
- Flood control;
- Mechanical equipment;
- Personal protective equipment;
- Explosives;

- Fire prevention and protection; and
- Emergency procedures, including evacuation plans and check-in/check-out systems.

When

There is no specific training time mentioned in the OSHA standard.

Self-Rescuers

The employer, having current approval from NIOSH and the Mine Safety and Health Administration (MSHA), shall provide training to self-rescuers to be immediately available to all employees at work stations in underground areas where employees might be trapped by smoke or gas. The selection, issuance, use, and care of respirators shall be in compliance with the respiratory protection standard (§1926.103 [b] and [c]).

Emergency Provisions

Rescue team members shall be qualified in rescue procedures, the use and limitations of a breathing apparatus, and the use of fire-fighting equipment. Qualifications shall be annually.

On job sites where flammable or noxious gases are encountered or anticipated in hazardous quantities, rescue team members shall practice donning and using a self-contained breathing apparatus monthly.

The employer shall ensure that rescue teams are familiar with conditions at the job site.

Air Quality and Monitoring

The employer shall assign a competent person who will perform all air monitoring required.

Monitoring of airborne contaminants will be "as often as necessary"; the competent person shall make a reasonable determination as to which substances to monitor and how frequently to monitor, taking into consideration the following:

- Location of job site;
- Geology of the job site;
- Presence of air contaminants in nearby job sites and changes in levels of substances monitored on prior shifts;
- Work practices and job-site conditions including use of diesel engines;
- Explosives;
- Fuel gas;
- Volume and flow of ventilation;
- Visible atmospheric conditions;
- Decompression of the atmosphere;

- Welding, cutting, and hot work; and
- Employees' physical reactions to working underground.

When the competent person determines, on the basis of air monitoring results or other information, that air contaminants might be present in sufficient quantity to be dangerous to life, the employer shall prominently post a notice at all entrances to the underground job site to inform all entrants of the hazardous condition and shall ensure that the necessary precautions are taken.

Ground Support (Underground Areas)

A competent person shall inspect the roof, face, and walls of the work area at the start of each shift and as often as necessary to determine ground stability.

A competent person shall determine whether rock bolts meet the necessary torque and determine the testing frequency in light of the bolt system, ground conditions, and the distance from vibration sources.

Hoisting Unique to Underground Construction

A competent person shall visually check all hoisting machinery, equipment, anchorages, and hoisting rope at the beginning of each shift during hoist use.

Each safety device shall be checked by a competent person at least weekly during hoist use to ensure suitable operation and safe condition.

DO EMPLOYEES OPERATE COMPRESSED AIR?

Then §1926.803 training applies:

Who

This training applies to all employees who operate compressed air. Every employee shall be instructed in the rules and regulations that concern his safety or the safety of others.

There shall be present, at all times, at least one competent person, designated by and representing the employer, who shall be familiar with this in all respects and be responsible for full compliance with these and other applicable subparts.

When

There shall be retained one or more licensed physicians familiar with and experienced in the physical requirements and the medical aspects of compressed air work and the treatment of decompression illness.

The physician shall be available at all times while work is in progress in order to provide medical supervision of employees employed in compressed air work. He or she shall be physically qualified and be willing to enter a pressurized environment.

Be in constant charge of an attendant under the direct control of the retained physician. The attendant shall be trained in the use of the lock and suitably instructed regarding steps to be taken in the treatment of employees exhibiting symptoms compatible with a diagnosis of decompression illness.

Every employee going under air pressure for the first time shall be instructed on how to avoid excessive discomfort.

In the event it is necessary for an employee to be in compressed air more than once in a 24-hour period, the appointed physician shall be responsible for the establishment of methods and procedures of decompression applicable to repetitive exposures.

If decanting is necessary, the appointed physician shall establish procedures before any employee is permitted to be decompressed by decanting methods.

The period of time employees spend at atmospheric pressure between the decompression following the shift and recompression shall not exceed five minutes.

At all times there shall be a thoroughly experienced, competent, and reliable person on duty at the air control valves as a gauge tender who shall regulate the pressure in the working areas. During tunneling operations, one gauge tender may regulate the pressure in not more than two headings, provided that the gauge and controls are all in one location. In caisson work, there shall be a gauge tender for each caisson.

Subpart T—Demolition

Table 5.16. Subpart T—Demolition training standards

Section	Designation
1926.850	Preparatory operations
1926.852	Chutes
1926.859	Mechanical demolition

The Subpart T standards are restricted in their application to employers engaged in demolition operations. There are three training requirements applicable to this subpart.

ARE EMPLOYEES INVOLVED IN DEMOLITION OPERATIONS?

Then §1926.850 training applies:

Who

This training applies to all employees involved in demolition operations.

When

Prior to permitting employees to start demolition operations, an engineering survey shall be made, by a competent person, of the structure to determine the condition of the framing, floors, and walls, and the possibility of an unplanned collapse of any portion of the structure. Any adjacent structure where employees may be exposed shall also be similarly checked. The employer shall have in writing evidence that such a survey has been performed.

CHUTES REQUIREMENTS

From §1926.852, a substantial gate shall be installed in each chute at or near the discharge end. A competent employee shall be assigned to control the operation of the gate and the backing and loading of trucks.

MECHANICAL DEMOLITION REQUIREMENTS

From §1926.859, during demolition, continuing inspections by a competent person shall be made as the work progresses to detect hazards resulting from weakened or deteriorated floors or walls, or from loosened material. No employee shall be permitted to work where hazards exist until they are corrected by shoring, bracing, or other effective means.

Subpart U—Blasting and Use of Explosives

Table 5.17. Subpart U—Blasting and use of explosives training standards

Section	Designation
1926.900	General provisions
1926.901	Blaster qualifications
1926.902	Surface transportation of explosives

Subpart U training requirements apply when employers use explosives or do blasting in their work. No employer should use explosives unless he or she is familiar with these requirements. There are three training requirements applicable to this subpart.

DO EMPLOYEES HANDLE OR USE EXPLOSIVES?

Then §1926.900 training applies:

Who

This training applies to all employees who handle and use explosives. The employer shall permit only authorized and qualified persons to handle and use explosives.

When

There must be a prominent display of adequate signs, warning against the use of mobile radio transmitters, on all roads within 1,000 feet of blasting operations. Whenever adherence to the 1,000-foot distance would create an operational handicap, a competent person shall be consulted to evaluate the particular situation, and alternative provisions may be made that are adequately designed to prevent any premature firing of electric blasting caps.

A description of any alternatives shall be reduced to writing and shall be certified as meeting the purposes of this subdivision by the competent person consulted. The description shall be maintained at the construction site during the duration of the work and shall be available for inspection by representatives of the secretary of labor.

All loading and firing shall be directed and supervised by competent persons thoroughly experienced in this field.

BLASTER TRAINING QUALIFICATIONS

From §1926.901, a blaster shall be qualified, by reason of training, knowledge, or experience, in the field of transporting, storing, handling, and using explosives, and have a working knowledge of state and local laws and regulations that pertain to explosives.

Blasters shall be required to show evidence of competency in handling explosives and performing in a safe manner the type of blasting that will be required.

The blaster shall be knowledgeable and competent in the use of each type of blasting method used.

SURFACE TRANSPORTATION OF EXPLOSIVES

From §1926.902, motor vehicles or conveyances transporting explosives shall only be driven by, and be in the charge of, a licensed driver who is physically fit. He or she shall be familiar with the local, state, and federal regulations governing the transportation of explosives.

Each vehicle used for transportation of explosives shall be equipped with a fully charged fire extinguisher, in good condition. An Underwriters Laboratory–approved extinguisher of not less than 10-ABC rating will meet the minimum requirement. The driver shall be trained in the use of the extinguisher on his vehicle.

Subpart V—Power Transmission and Distribution

Table 5.18. Subpart V—Power transmission and distribution training standards

Section	Designation
1926.950	General requirements
1926.955	Overhead lines
1926.956	Underground lines
1926.957	Construction in energized substations

The Subpart V training standards are limited in their application to construction, erection, alteration, conversion, and improvement of electric transmission and distribution lines and equipment. There are four training requirements applicable to this subpart.

ARE EMPLOYEES INVOLVED IN POWER TRANSMISSION AND DISTRIBUTION?

Then §1926.950 training applies:

Who

This training applies to all employees who are involved in transmission and distribution lines that include overhead lines, underground lines, and construction in energized substations.

When

There is no specific training time mentioned in the OSHA standard.

When de-energizing lines and equipment operate in excess of 600 volts, and the means of disconnecting from electric energy is not visibly open or visibly locked out, the following provisions of this training requirement shall be complied with:

- Notification and assurance from the designated employee, a qualified person delegated to perform specific duties under the existing conditions, shall be obtained.
- All switches and disconnectors through which electric energy might be supplied to the particular section of line or equipment to be worked have been de-energized.
- All switches and disconnectors are plainly tagged indicating that men are at work.
- And where design of switches and disconnectors permit, they have been rendered inoperable.

When more than one independent crew requires the same line or equipment to be de-energized, a prominent tag for each independent crew shall be placed on the line or equipment by the designated employee in charge.

Upon completion of work on de-energized lines or equipment, each designated employee in charge shall determine that all employees in his or her crew are clear, determine that protective grounds installed by his crew have been removed, and report to the designated authority that all tags protecting the crew may be removed.

When a crew working on a line or equipment can clearly see that the means of disconnecting from electric energy are visibly open or visibly locked out, the provisions of this paragraph shall apply.

Upon completion of work on de-energized lines or equipment, each designated employee in charge shall determine that all employees in his crew are clear, determine that protective grounds installed by his crew have been removed, and report to the designated authority that all tags protecting his crew may be removed.

The employer shall provide training or require that his employees are knowledgeable and proficient in procedures involving emergency situations and first aid fundamentals including resuscitation.

ARE EMPLOYEES EXPOSED TO OVERHEAD LINES?

Then §1926.955 training applies:

Who

Supervisors of live-line, bare-hand work are to be trained and qualified. Employees using the live-line, bare-hand technique on energized circuits are to be trained.

When

Prior to stringing parallel to an existing energized transmission line, a competent determination shall be made to ascertain whether dangerous induced voltage buildups will occur, particularly during switching and ground-fault conditions.

Employees shall be instructed and trained in the live-line, bare-hand technique and the safety requirements pertinent before being permitted to use the technique on energized circuits.

All work shall be personally supervised by a person trained and qualified to perform live-line, bare-hand work.

ARE EMPLOYEES EXPOSED TO UNDERGROUND LINES?

Then §1926.956 training applies:

Who

While work is being performed in manholes, an employee shall be available in the immediate vicinity to render emergency assistance as might be required. This shall not preclude the employee in the immediate vicinity from occasionally entering a manhole to provide assistance, other than emergency.

When

This requirement does not preclude a qualified employee (a person who by reason of experience or training is familiar with the operation to be performed and the hazards involved), working alone, from entering for brief periods of time a manhole where energized cables or equipment are in service, for the purpose of inspection, housekeeping, taking readings, or similar work that can be performed safely.

DO EMPLOYEES PERFORM CONSTRUCTION WORK IN ENERGIZED SUBSTATIONS?

Then §1926.957 training applies:

Who

When construction work is performed in an energized substation, authorization shall be obtained from the designated, authorized person (a qualified person delegated to perform specific duties).

When

Under the conditions existing before work is started, work on or adjacent to energized control panels shall be performed by designated employees.

Use of vehicles, gin poles, cranes, and other equipment in restricted or hazardous areas shall at all times be controlled by designated employees.

Subpart X—Stairways and Ladders

Table 5.19.   Subpart X—Stairways and ladders training standards

Section	Designation
1926.1060	Training requirements

The Subpart X training standards set forth the conditions and circumstances under which ladders and stairways must be provided and used. The standard applies to all stairways and ladders used in construction, alteration, repair, demolition, painting, and repair work.

Any employer who has one or more employees engaged in construction work who will use a ladder or stairway while working (and that covers the vast majority of construction employers) must become familiar with the Subpart X requirements.

OSHA standard §1926.1060 requires that a training program be provided for each employee using ladders and stairways so that he or she will be able to recognize the hazards related to ladders and stairways and know the procedures to be followed in order to minimize those hazards.

DO EMPLOYEES USE LADDERS OR STAIRWAYS?

Then §1926.1060 training applies:

Who

This training applies to all employees who use ladders and stairways. Ladders shall be inspected by a competent person for visible defects on a periodic basis and after any occurrence that could affect their safe use.

When

The employer shall provide a training program for each employee using ladders and stairways, as necessary. The program shall enable each employee to recognize hazards related to ladders and stairways, and shall train each employee in the procedures to be followed to minimize these hazards.

The employer shall ensure that each employee has been trained by a competent person in the following areas, as applicable:

- The nature of fall hazards in the work area;
- The correct procedures for erecting, maintaining, and disassembling the fall protection systems to be used;
- The proper construction, use, placement, and care in handling of all stairways and ladders; and
- The maximum intended load-carrying capacities of ladders used.

Retraining shall be provided for each employee as necessary so that the employee maintains the understanding and knowledge acquired through compliance.

Subpart Y—Commercial Diving Operations

Table 5.20. Subpart Y—Commercial diving operations training standards

Section	Designation
1926.1076	Qualifications of dive team

Subpart Y applies to commercial diving operations. It contains only one training standard that concerns diving and related support operations in connection with all types of commercial diving work.

ARE EMPLOYEES INVOLVED IN COMMERCIAL DIVING OPERATIONS?

Then §1926.1076 training applies:

Who

This training applies to all employees who are involved in commercial diving operations. Each dive team member shall have the experience or training necessary to perform assigned tasks in a safe and healthful manner.

Each dive team member shall have the experience or training in the following:

- The use of tools, equipment, and systems relevant to assigned tasks;
- Techniques of the assigned diving mode; and
- Diving operations and emergency procedures.

All dive team members shall be trained in cardiopulmonary resuscitation and first aid (American Red Cross standard course or equivalent).

Dive team members who are exposed to or control the exposure of others to hyperbaric conditions shall be trained in diving-related physics and physiology.

When

Each dive team member shall be assigned tasks in accordance with the employee's experience or training, except that limited additional tasks may be assigned to employees undergoing training provided that these tasks are performed under the direct supervision of an experienced dive team member.

Designated Person-in-Charge Training

The designated person in charge shall have experience and training in the conduct of the assigned diving operation.

Subpart Z—Toxic and Hazardous Substances

Table 5.20. Subpart Z—Toxic and hazardous substances training standards

Section	Designation
§1926.1101	Asbestos
§1926.1102	Coal tar pitch volatiles: interpretation of term
§1926.1103	13 carcinogens (4-Nitrobiophenyl, etc.)
§1926.1104	alpha-Naphthylamine
§1926.1106	Methyl chloromethyl ether
§1926.1107	3.3'-Dichlorobenzidine (and its salts)
§1926.1108	bis-Chloromethyl ether
§1926.1109	beta-Naphthylamine
§1926.1110	Benzidine
§1926.1111	4-Aminodiphenyl
§1926.1112	Ethyleneimine
§1926.1113	beta-Propiolactone
§1926.1114	2-Acetylaminofluorene
§1926.1115	4-Dimethylaminoazobenzene
§1926.1116	N-Nitrosodimethylamine
§1926.1117	Vinyl chloride
§1926.1118	Inorganic arsenic
§1926.1126	Hexavalent chromium, Cr(VI)
§1926.1127	Cadmium
§1926.1128	Benzene
§1926.1129	Coke oven emissions
§1926.1144	1,2-dibromo-3-chloropropane
§1926.1145	Acrylonitrile
§1926.1147	Ethylene oxide
§1926.1148	Formaldehyde
§1926.1152	Methylene chloride

The Subpart Z training standards contain various substance-specific requirements that set airborne permissible exposure limits (PELs) for over 450 listed substances. A PEL is the maximum amount of a contaminant in the air to which workers may be exposed over a given period of time.

They impose a number of specific requirements to be implemented in workplaces where those substances are present and require employers, chemical manufacturers, and importers to provide information upon all hazardous chemicals through a variety of methods.

ARE EMPLOYEES EXPOSED TO ASBESTOS?

Then §1926.1101 training applies:

Who

The employer shall provide a training program to all employees who are involved with installation, removal, repair, and maintenance of certain roofing and pipeline coating materials.

Before work begins and as needed during the job, a competent person who is capable of identifying asbestos hazards in the workplace and selecting the appropriate control strategy for asbestos exposure, and who has the authority to take prompt corrective measures to eliminate hazards, shall conduct an inspection of the worksite to determine if the roofing material is intact and will likely remain intact.

All employees performing work shall be trained in a training program that meets the requirements of this section.

Respiratory Protection Training

For employees who use respirators, the employer must provide respirators. Respirators must be used during the following:

- Class I asbestos work;
- Class II asbestos work when asbestos-containing material (ACM) is not removed in a substantially intact state;
- Class II and III asbestos work that is not performed using wet methods, except for removal of ACM from sloped roofs when a negative-exposure assessment has been conducted and ACM is removed in an intact state;
- Class II and III asbestos work for which a negative-exposure assessment has not been conducted;
- Class III asbestos work when thermal system insulation (TSI) or surfacing ACM or presumed asbestos-containing material (PACM) is being disturbed;
- Class IV asbestos work performed within regulated areas where employees who are performing other work are required to use respirators; and
- Work operations covered by this section for which employees are exposed above the time-weighted average (TWA) or excursion limit.

When

The employer shall at no cost to the employee institute a training program for all employees who are likely to be exposed in excess of a PEL and for all employees who perform Class I through IV asbestos operations, and shall ensure their participation in the program.

Training shall be provided prior to or at the time of initial assignment and at least annually.

Training for Class I Operations and for Class II Operations

The use of critical barriers (or equivalent isolation methods) or negative pressure enclosures under this section shall be the equivalent in curriculum, training method, and length to the Environmental Protection Agency (EPA) Model Accreditation Plan's (MAP) asbestos-abatement workers training.

For work with asbestos-containing roofing materials, flooring materials, siding materials, ceiling tiles, or transite panels, training shall include specific work practices and engineering controls of this section that specifically relate to that category. Such course shall include "hands-on" training and shall take at least eight hours.

An employee who works with more than one of the categories of materials specified in this section shall receive training in the work practices applicable to each category of material that the employee removes and each removal method that the employee uses.

Training for Class II Operations

Employers shall provide all the specific work practices and engineering controls that specifically relate to the category of material being removed and shall include "hands-on" training in the work practices applicable to each category of material that the employee removes and each removal method that the employee uses.

Training for Class III Employees

Employers shall be consistent with EPA requirements for training of local education agency maintenance and custodial staff. Such a course shall also include "hands-on" training and shall take at least 16 hours. One exception is that, for Class III operations for which the competent person determines that the EPA curriculum does not adequately cover the training needed to perform that activity, training shall include in addition the specific work practices and engineering controls of this section that specifically relate to that activity, and shall include "hands-on" training in the work practices applicable to each category of material that the employee disturbs.

Training for Employees Performing Class IV Operations

Employers shall be consistent with EPA requirements for training of local education agency maintenance and custodial staff. Such a course shall include available information concerning the locations of thermal system insulation and surfacing

ACM/PACM and asbestos-containing flooring material, or flooring material where the absence of asbestos has not yet been certified, and instruction in recognition of damage, deterioration, and delamination of asbestos-containing building materials. Such a course shall take at least two hours.

The training program shall be conducted in a manner so that the employee is able to understand, and the employer shall ensure that each such employee is informed of the following:

- Methods of recognizing asbestos, including the requirement of this section to presume that certain building materials contain asbestos;
- The health effects associated with asbestos exposure;
- The relationship between smoking and asbestos in producing lung cancer;
- The nature of operations that could result in exposure to asbestos;
- The importance of necessary protective controls to minimize exposure including, as applicable, engineering controls, work practices, respirators, housekeeping procedures, hygiene facilities, protective clothing, decontamination procedures, emergency procedures, and waste disposal procedures, and any necessary instruction in the use of these controls and procedures;
- Where Class III and IV work will be or is performed;
- The purpose, proper use, fitting instructions, and limitations of respirators as required by 29 CFR §1910.134;
- The appropriate work practices for performing the asbestos job;
- Medical surveillance program requirements;
- The content of this standard including appendixes;
- The names, addresses, and phone numbers of public health organizations that provide information and materials or conduct programs concerning smoking cessation (the employer may distribute the list of such organizations contained in appendix J to this section, to comply with this requirement); and
- The requirements for posting signs and affixing labels and the meaning of the required legends for such signs and labels.

Training Materials

The employer shall make readily available to affected employees, without cost, written materials relating to the employee training program, including a copy of this regulation.

The employer shall provide to the assistant secretary and the director, upon request, all information and training materials relating to the employee information and training program.

Competent Person

On all construction worksites covered by this standard, the employer shall designate a competent person, having the qualifications and authorities for ensuring worker safety and health.

Inspections must be made by the competent person and require health and safety prevention programs to provide for frequent and regular inspections of the job sites, materials, and equipment.

In addition, the competent person shall make frequent and regular inspections of the job sites. For Class I jobs, onsite inspections shall be made at least once during each work shift and at any time at employee request. For Class II, III, and IV jobs, onsite inspections shall be made at intervals sufficient to assess whether conditions have changed and at any reasonable time at employee request.

Training for the Competent Person

For Class I and II asbestos work the competent person shall be trained in all aspects of asbestos removal and handling, including abatement, installation, removal, and handling; the contents of this standard; the identification of asbestos; removal procedures, where appropriate; and other practices for reducing the hazard. Such training shall be obtained in a comprehensive course for supervisors that meets the criteria of the EPA's Model Accredited Plan, such as a course conducted by an EPA-approved or state-approved training provider, certified by the EPA or a state, or a course equivalent in stringency, content, and length.

For Class III and IV asbestos work, the competent person shall be trained in aspects of asbestos handling appropriate for the nature of the work, to include procedures for setting up glove bags and minienclosures, practices for reducing asbestos exposures, use of wet methods, the contents of this standard, and the identification of asbestos. Training shall include successful completion of a course that is consistent with EPA requirements for training of local education agency maintenance and custodial staff, or its equivalent in stringency, content, and length.

ARE EMPLOYEES EXPOSED TO ANY OF THE 13 TOXIC AND HAZARDOUS SUBSTANCES LISTED BELOW?

- 4-Nitrobiphenyl—§1926.1103
- Alpha-Naphthylamine—§1926.1104
- Methyl chloromethyl ether—§1926.1106
- 3,3'-Dichlorobenzidine (and its salts)—§1926.1107
- Bis-Chloromethyl ether—§1926.1108
- Beta-Naphthylamine—§1926.1109
- Benzidine—§1926.1110
- 4-Aminodiphenyl—§1926.1111
- Ethyleneimine—§1926.1112
- Beta-Propiolactone—§1926.1113
- 2-Acetylaminofluorene—§1926.1114
- 4 Dimethylaminoazobenzene—§1926.1115
- N-Nitrosodimethylamine—§1926.1116

Then training applies:

Who

This training applies to all authorized employees who perform work where they may be exposed to the 13 carcinogens that are manufactured, processed, repackaged, released, handled, or stored but shall not apply to transshipment in sealed containers.

Authorized employees means those employees assigned to work where a regulated chemical is manufactured, processed, used, repackaged, released, handled, or stored.

When

Each employee prior to being authorized to enter a regulated area shall receive a training and indoctrination program including the following:

- The nature of the carcinogenic hazards of a carcinogen addressed by this section, including local and systemic toxicity;
- The specific nature of the operation involving the listed substances that could result in exposure;
- The purpose for and application of the medical surveillance program, including, as appropriate, methods of self-examination;
- The purpose for an application of decontamination practices and procedures;
- The purpose for and significance of emergency practices and procedures;
- The employee's specific role in emergency procedures;
- Specific information to aid the employee in recognition and evaluation of conditions and situations that might result in the release of 4-Nitrobiophenyl;
- The purpose for and application of specific first aid procedures and practices; and
- A review of the training section at the employee's first training and indoctrination program and annually thereafter.

Specific emergency procedures shall be prescribed and posted, and employees shall be familiarized with their terms and rehearsed in their application.

ARE EMPLOYEES EXPOSED TO VINYL CHLORIDE?

Then §1926.1117 training applies:

Who

Each employee engaged in vinyl chloride or polyvinyl chloride operations shall be provided training in a program relating to the hazards of vinyl chloride and precautions for its safe use. The program should include the following:

- The nature of the health hazard from chronic exposure to vinyl chloride including, specifically, the carcinogenic hazard;
- The specific nature of operations that could result in exposure to vinyl chloride in excess of the permissible limit and necessary protective steps;

- The purpose for, proper use of, and limitations of respiratory protective devices;
- The fire hazard and acute toxicity of vinyl chloride and the necessary protective steps;
- The purpose for and a description of the monitoring program;
- The purpose for and a description of the medical surveillance program; and
- Specific information to aid the employee in recognition of conditions that might result in the release of vinyl chloride.

When

Employees must be trained at the first training and indoctrination program, and annually thereafter.

ARE EMPLOYEES EXPOSED TO INORGANIC ARSENIC?

Then §1926.1118 training applies:

Who

This training applies to all employees who are subject to exposure to inorganic arsenic above the action level without regard to respirator use, or for whom there is a possibility of skin or eye irritation from inorganic arsenic. The employer shall ensure that those employees participate in the training program.

When

The training program shall be provided for all employees at the time of initial assignment for those subsequently covered by this standard, and shall be repeated at least quarterly for employees who have optional use of respirators and at least annually for other covered employees thereafter; the employer shall ensure that each employee is informed of the following:

- The quantity, location, manner of use, storage, sources of exposure, and the specific nature of operations that could result in exposure to inorganic arsenic as well as any necessary protective steps;
- The purpose, proper use, and limitation of respirators;
- The purpose and a description of the medical surveillance program; and
- The engineering controls and work practices associated with the employee's job assignment.

The employer shall make available to all affected employees a copy of this standard and its appendixes and provide upon request all materials relating to the employee information and training program to the assistant secretary and the director.

ARE EMPLOYEES EXPOSED TO CADMIUM?

Then §1926.1127 training applies:

Who

The employer shall institute a training program for all employees who are potentially exposed to cadmium, ensure employee participation in the program, and maintain a record of the contents of the program.

When

Training shall be provided prior to or at the time of initial assignment to a job involving potential exposure to cadmium and at least annually. The employer shall make the training program understandable to the employee and ensure that each employee is informed of the following:

- The health hazards associated with cadmium exposure, with special attention to the information incorporated in appendix A of the standard;
- The quantity, location, manner of use, release, and storage of cadmium in the workplace and the specific nature of operations that could result in exposure to cadmium, especially exposures above the permissible exposure limits (PELs);
- The engineering and work practices associated with the employee's job assignment;
- The measures employees can take to protect themselves from exposure to cadmium, including modification of such habits as smoking and personal hygiene, and specific procedures the employer has implemented to protect employees from exposure to cadmium such as appropriate work practices, emergency procedures, and the provision of personal protective equipment;
- The purpose, proper selection, fitting, proper use, and limitations of respirators and protective clothing; and
- The purpose and a description of the medical surveillance programs training section.

The employer shall make a copy of this training section and its appendixes readily available without cost to all affected employees and shall provide a copy if requested.

ARE EMPLOYEES EXPOSED TO BENZENE?

Then §1926.1128 training applies:

Who

This training applies to all employees who are exposed to benzene.

When

The employer shall provide employees with information and training at the time of their initial assignment to a work area where benzene is present. If exposures are above the action level, employees shall be provided with information and training at least annually thereafter.

The training program shall be in accordance with the requirements of the hazard communication program and include specific information on benzene for each category of information included.

The employer shall also provide employees with an explanation of the contents of this training section, including appendixes A and B; indicate to them where the standard is available; and describe the medical surveillance program.

ARE EMPLOYEES EXPOSED TO COKE OVEN EMISSIONS?

Then §1926.1129 training applies:

Who

The employer shall institute a training program for all employees exposed to coke oven emissions in the regulated area and shall ensure their participation.

When

The training program shall be provided at the time of initial assignment and at least annually for all employees who are employed in the regulated area, except that training regarding the occupational safety and health hazards associated with exposure to coke oven emissions and the purpose, proper use, and limitations of respiratory protective devices shall be provided at least quarterly. The training program shall include informing each employee of the following:

- The information contained in the substance information sheet for coke oven emissions;
- The purpose, proper use, and limitations of respiratory protective devices; and
- The purpose for and a description of the medical surveillance program including information on the occupational safety and health hazards associated with exposure to coke oven emissions.

The employer shall make a copy of this standard and its appendixes readily available to all employees who are employed in the regulated area.

ARE EMPLOYEES EXPOSED TO 1,2-DIBROMO-3-CHLOROPROPANE?

Then §1926.1144 training applies:

Who

The employer shall institute a training program for all employees who might be exposed to 1,2-dibromo-3-chloropropane (DBCP) and shall ensure their participation in a training program. The employer shall ensure that each employee is informed of the following:

- The quantity, location, manner of use, release or storage of DBCP and the specific nature of operations that could result in exposure to DBCP as well as any necessary protective steps;
- The purpose, proper use, and limitations of respirators; and
- The purpose and description of the medical surveillance program.

When

The employer shall make a copy of this standard and its appendixes available to all affected employees. The employer shall provide, upon request, all materials relating to the employee information and training program to the assistant secretary and the director.

ARE EMPLOYEES EXPOSED TO ACRYLONITRILE?

Then §1926.1145 training applies:

Who

The employer shall institute a training program for and ensure the participation of all employees exposed to acrylonitrile (AN) above the action level, all employees whose exposures are maintained below the action level by engineering and work practice controls, and all employees subject to potential skin or eye contact with liquid AN.

When

Training shall be provided at the time of initial assignment, or upon institution of the training program, and at least annually thereafter, and the employer shall ensure that each employee is informed of the following:

- The quantity, location, manner of use, release, or storage of AN and the specific nature of operations that could result in exposure to AN, as well as any necessary protective steps;
- The purpose, proper use, and limitations of respirators and protective clothing;
- The purpose and a description of the medical surveillance program;
- The emergency procedures developed; and
- Engineering and work practice controls, their function, and the employee's relationship to these controls.

The employer shall make a copy of this standard and its appendixes readily available to all affected employees.

ARE EMPLOYEES EXPOSED TO ETHYLENE OXIDE?

Then §1926.1147 training applies:

Who

This training applies to all employees who might be exposed to ethylene oxide (EtO), and they shall be informed of the following:

- Any operations in their work area where EtO is present;
- The location and availability of the written EtO final rule; and
- The medical surveillance program.

When

The employer shall provide employees who are potentially exposed to EtO at or above the action level with information and training on EtO at the time of initial assignment and at least annually thereafter.

Employer training shall include

- methods and observations that may be used to detect the presence or release of EtO in the work area (such as monitoring conducted by the employer, continuous monitoring devices, etc.);
- the physical and health hazards of EtO;
- the measures employees can take to protect themselves from hazards associated with EtO exposure, including specific procedures the employer has implemented to protect employees from exposure to EtO, such as work practices, emergency procedures, and personal protective equipment to be used; and
- the details of the hazard communication program developed by the employer, including an explanation of the labeling systems and how employees can obtain and use the appropriate hazard information.

The details of the hazard communication program developed by the employer, including an explanation of the labeling systems and how employees can obtain and use the appropriate hazard information, must be provided to employees.

ARE EMPLOYEES EXPOSED TO FORMALDEHYDE?

Then §1926.1148 training applies:

Who

This training applies to all employees who are assigned to workplaces where there is exposure to formaldehyde, except where the employer can show, using objective data, that employees are not exposed to formaldehyde above 0.1 ppm (in this case, the employer is not required to provide training).

When

Employers shall provide information and training to employees at the time of initial assignment and whenever a new exposure to formaldehyde is introduced into the work. The training shall be repeated at least annually.

The training program shall be conducted in a manner in which the employee is able to understand and shall include the following:

- A discussion of the contents of this regulation and the contents of the material safety data sheet;
- The purpose for and a description of the medical surveillance program;
- A description of the potential health hazards associated with exposure to formaldehyde and a description of the signs and symptoms of exposure to formaldehyde;
- Instructions to immediately report to the employer the development of any adverse signs or symptoms that the employee suspects is attributable to formaldehyde exposure;
- Description of operations in the work area where formaldehyde is present and an explanation of the safe work practices appropriate for limiting exposure to formaldehyde in each job;
- The purpose for, proper use of, and limitations of personal protective clothing and equipment;
- Instructions for the handling of spills, emergencies, and cleanup procedures;
- An explanation of the importance of engineering and work practice controls for employee protection and any necessary instruction in the use of these controls; and
- A review of emergency procedures including the specific duties or assignments of each employee in the event of an emergency.

ARE EMPLOYEES EXPOSED TO METHYLENE CHLORIDE?

Then §1926.1152 training applies:

Who

The training applies to all employees who have exposure to airborne concentrations of methylene chloride (MC) that exceeds or can reasonably be expected to exceed the action level. The employer shall inform each affected employee of the quantity, location, manner of use, release, and storage of MC and the specific operations in the workplace that could result in exposure to MC, particularly noting where exposures might be above the eight-hour TWA PEL or short-term exposure limit (STEL).

This program also applies to retraining affected employees as necessary to ensure that each employee exposed above the action level or the STEL maintains the requisite understanding of the principles of safe use and handling of MC in the workplace.

When

The employer shall provide information and training for each affected employee prior to or at the time of initial assignment to a job involving potential exposure to MC. The employer shall ensure that information and training is presented in a manner that is understandable to the employees.

The employer shall inform each affected employee of the requirements of this section and information available in its appendixes, as well as how to access or obtain a copy of it in the workplace.

The employer shall retrain each affected employee as necessary to ensure that each employee exposed above the action level or the STEL maintains the requisite understanding of the principles of safe use and handling of MC in the workplace.

Whenever there are workplace changes, such as modifications of tasks or procedures or the institution of new tasks or procedures, which increase employee exposure, and where those exposures exceed or can reasonably be expected to exceed the action level, the employer shall update the training as necessary to ensure that each affected employee has the requisite proficiency.

An employer whose employees are exposed to MC at a multiemployer worksite shall notify the other employers with work operations at that site in accordance with the requirements of the Hazard communication standard, 29 CFR §1910.1200, as appropriate.

CHAPTER 6

OSHA's Construction Checklists Simplified

The objective of this checklist is to assist employers with evaluating safety and health on the job site. The standards referred to are federal occupational safety and health standards for the construction industry, 29 CFR 1926.

The checklist is designed in such a manner that a negative answer to any question indicates an area of concern. The checklist is intended as a guide and does not necessarily address all construction-related standards. You are encouraged to incorporate this checklist in the needs of your organization. Feel free to add to it or redevelop it in a manner that will meet your specific needs.

Note that this construction safety inspection checklist is not designed to supersede existing safety inspection checklists; rather, it should be used only as a general guideline. You are encouraged to customize this general guideline to accommodate your specific operations.

Construction Safety Inspection Checklist (1926)

Company name: _____

Job site address: _____

Superintendent: _____

Date/time: _____

Inspector(s): _____

Please check items inspected with the following:

    Yes   No   N/A   Date corrected

Administrative Requirements (1904)

1. Is the U.S. Department of Labor job safety and health protection poster (or a facsimile) posted in a conspicuous place?

    Yes _____ No _____ N/A _____ Date corrected _____

2. Are all occupational deaths, injuries, and illnesses recorded on the Occupational Safety and Health Administration (OSHA) Form 300 as required?
Yes _____ No _____ N/A _____ Date corrected _____

3. Is each recordable injury entered on the OSHA 300, Log of Work-Related Injuries and Illnesses, within seven calendar days of receiving information that a recordable injury or illness has occurred?
Yes _____ No _____ N/A _____ Date corrected _____

4. Is the OSHA 300-A, Summary of Work-Related Injuries and Illnesses, for the previous year posted from February 1 to April 30?
Yes _____ No _____ N/A _____ Date corrected _____

5. Is the OSHA form 301 (or its equivalent), a supplemental record of each occupational injury or illness, available?
Yes _____ No _____ N/A _____ Date corrected _____

6. Is the OSHA area director notified within eight hours of any employment fatality or any accident that results in the hospitalization of three or more employees?
Yes _____ No _____ N/A _____ Date corrected _____

Subpart C—General Safety and Health Provisions

GENERAL SAFETY AND HEALTH—1926.20

1. Is each employee instructed in the recognition and avoidance of unsafe conditions and the regulations applicable to his or her work environment to control or eliminate any hazards or other exposure to illness or injury?
Yes _____ No _____ N/A _____ Date corrected _____

2. Are employees who are required to handle or use poisons, caustics, and other harmful substances instructed in their safe handling and use, and made aware of the potential hazards, personal hygiene, and personal protective measures?
Yes _____ No _____ N/A _____ Date corrected _____

3. Are employees who are required to enter confined or enclosed spaces instructed as to the nature of the hazards involved, the necessary precautions to be taken, and in the use of protective and emergency equipment?
Yes _____ No _____ N/A _____ Date corrected _____

4. Is form and scrap lumber with protruding nails and all other debris kept cleared from work areas, passageways, and stairs?
Yes _____ No _____ N/A _____ Date corrected _____

5. Are employees required to wear appropriate personal protective equipment when there is an exposure to hazardous conditions?
Yes _____ No _____ N/A _____ Date corrected _____

Subpart D—Occupational Health and Environmental Controls

MEDICAL SERVICES, FIRST AID, AND SANITATION—1926.50

1. Is a facility for the treatment of injured employees located within three to five minutes of the job site? If not, is there an employee trained in first aid at the site?
   Yes _____ No _____ N/A _____ Date corrected _____
2. Are first aid supplies readily accessible?
   Yes _____ No _____ N/A _____ Date corrected _____
3. Are telephone numbers of physicians, hospitals, or ambulances conspicuously posted?
   Yes _____ No _____ N/A _____ Date corrected _____
4. Are potable (drinking) water and adequate toilet facilities available at the job site?
   Yes _____ No _____ N/A _____ Date corrected _____
5. Are the regulations concerning protection of employees against the effects of noise exposure understood and complied with?
   Yes _____ No _____ N/A _____ Date corrected _____

HAZARD COMMUNICATION—1926.59

1. Does the employer have a written hazard communication program?
   Yes _____ No _____ N/A _____ Date corrected _____
2. Does the employer have a complete list of hazardous chemicals used on site and list the appropriate material safety data sheets (MSDSs)?
   Yes _____ No _____ N/A _____ Date corrected _____
3. Does the employer provide other employees with MSDSs or make MSDSs available at a central worksite location?
   Yes _____ No _____ N/A _____ Date corrected _____
4. Does the employer inform other employers of any precautionary measures they might need to take?
   Yes _____ No _____ N/A _____ Date corrected _____
5. Does the employer inform other employers of the labeling system?
   Yes _____ No _____ N/A _____ Date corrected _____
6. Are containers of hazardous chemicals labeled, tagged, or marked?
   Yes _____ No _____ N/A _____ Date corrected _____
7. Does labeling include both identities of the chemical and appropriate hazard warnings?
   Yes _____ No _____ N/A _____ Date corrected _____
8. Does the employer have an MSDS for each hazardous chemical on site?
   Yes _____ No _____ N/A _____ Date corrected _____

9. Are MSDSs available to all employees?
   Yes _____ No _____ N/A _____ Date corrected _____

10. Are employees trained on the hazardous chemicals in their work area?
    Yes _____ No _____ N/A _____ Date corrected _____

11. Does hazard communication training include the following?

- Any operation in employee's area where hazardous chemicals may be present?
  Yes _____ No _____ N/A _____ Date corrected _____
- Location and availability of the hazard communication program?
  Yes _____ No _____ N/A _____ Date corrected _____
- Methods that may be used to detect a chemical release?
  Yes _____ No _____ N/A _____ Date corrected _____
- Physical and chemical hazards of chemicals in the workplace?
  Yes _____ No _____ N/A _____ Date corrected _____
- Measures employees can take to protect themselves?
  Yes _____ No _____ N/A _____ Date corrected _____
- Details of the employer's hazard communication program (labeling and MSDSs)?
  Yes _____ No _____ N/A _____ Date corrected _____

12. Does the employer have a method of informing employees of the hazards of non-routine tasks, unlabeled pipes, and so forth?
    Yes _____ No _____ N/A _____ Date corrected _____

Subpart E—Personal Protective and Life Saving Equipment

PERSONAL PROTECTIVE EQUIPMENT—1926.100

1. Are protective helmets (hard hats) worn at all times where there is a possible danger of head injury from impact, falling or flying objects, or electrical shock and burns?
   Yes _____ No _____ N/A _____ Date corrected _____

2. Are employees provided with eye and face protection when operations present potential eye or face injury?
   Yes _____ No _____ N/A _____ Date corrected _____

3. Are safety nets provided when work areas are more than 25 feet above ground or water surfaces and when the use of ladders, scaffolds, catch platforms, temporary floors, safety lines, or safety belts are not practical?
   Yes _____ No _____ N/A _____ Date corrected _____

Subpart F—Fire Protection and Prevention

FIRE PROTECTION AND PREVENTION—1926.150

1. Has a fire protection program been developed?
   Yes _____ No _____ N/A _____ Date corrected _____

2. Is fire-fighting equipment conspicuously located and accessible?
    Yes _____ No _____ N/A _____ Date corrected _____
3. Is fire-fighting equipment periodically inspected and maintained in operating condition?
    Yes _____ No _____ N/A _____ Date corrected _____
4. Is fire-fighting equipment selected and provided according to the listed requirements?
    Yes _____ No _____ N/A _____ Date corrected _____
5. Are fire extinguishers located in such a manner that travel distance does not exceed 100 feet?
    Yes _____ No _____ N/A _____ Date corrected _____

FLAMMABLE AND COMBUSTIBLE LIQUIDS—1926.152

1. Are all flammable and combustible liquids stored and handled in approved containers and portable tanks?
    Yes _____ No _____ N/A _____ Date corrected _____
2. If more than 25 gallons of flammable or combustible liquid is stored in a room, is it in an approved cabinet?
    Yes _____ No _____ N/A _____ Date corrected _____
3. Does each service and fueling area have at least one portable fire extinguisher with a rating of not less than 20-B:C located within 75 feet of each pump, dispenser, underground file pipe opening, and lubrication or services area?
    Yes _____ No _____ N/A _____ Date corrected _____

Subpart G—Signs, Signals, and Barricades

SIGNS, SIGNALS, AND BARRICADES—1926.200

1. Are required signs, symbols, and accident-prevention tags being used in the workplace where appropriate?
    Yes _____ No _____ N/A _____ Date corrected _____
2. Are flagmen equipped with flags (at least 18 inches square), sign paddles, or lights?
    Yes _____ No _____ N/A _____ Date corrected _____
3. Are flagmen wearing red or orange warning garments?
    Yes _____ No _____ N/A _____ Date corrected _____
4. If working at night, are warning garments reflectorized?
    Yes _____ No _____ N/A _____ Date corrected _____

Subpart H—Materials Handling, Storage, Use, and Disposal

MATERIALS HANDLING, STORAGE, USE, AND DISPOSAL—1926.250

1. Are materials that are stored in tiers either stacked, racked, blocked, interlocked, or otherwise secured to prevent sliding, falling, or collapse?
   Yes _____ No _____ N/A _____ Date corrected _____

2. Are materials stored more than six feet from any hoistway or inside floor opening and more than ten feet from any exterior walls that do not extend above the top of the stored materials?
   Yes _____ No _____ N/A _____ Date corrected _____

3. Are aisles and passageways kept clear and in good repair?
   Yes _____ No _____ N/A _____ Date corrected _____

4. Are waste materials disposed of properly?
   Yes _____ No _____ N/A _____ Date corrected _____

5. Do alloy steel chain slings have a permanently affixed durable identification stating size, grade, capacity, and manufacturer?
   Yes _____ No _____ N/A _____ Date corrected _____

6. Do any hooks, rings, oblong links, pear-shaped links, coupling links, and other attachments have a rated capacity at least that of the chain?
   Yes _____ No _____ N/A _____ Date corrected _____

7. Is all rigging equipment for material handling inspected prior to use on each shift?
   Yes _____ No _____ N/A _____ Date corrected _____

8. When forming eyes in wire rope are U-bolt clips properly spaced and installed?
   Yes _____ No _____ N/A _____ Date corrected _____

Subpart I—Tools (Hand and Power)

TOOLS (HAND AND POWER)—1926.300

1. Are hand and power tools furnished by the employers or employees maintained in a safe condition?
   Yes _____ No _____ N/A _____ Date corrected _____

2. Are power tools, belts, gears, shafts, pulleys, sprockets, spindles, drums, fly wheels, and chains properly guarded?
   Yes _____ No _____ N/A _____ Date corrected _____

3. Are electric power operation tools and equipment properly grounded or double insulated?
   Yes _____ No _____ N/A _____ Date corrected _____

4. Are all employees who operate powder-actuated tools trained in the use of the particular tool they use?
   Yes _____ No _____ N/A _____ Date corrected _____

5. Do all circular saws have an exhaust hood or a guard to prevent accidental contact with the saw blade if there is a possibility of contact either beneath or behind the table?
   Yes _____ No _____ N/A _____ Date corrected _____

6. Do all portable circular saws have a guard above the base plate and a guard below the base plate that will automatically and instantly return to the covering position when the saw is withdrawn from the work?
   Yes _____ No _____ N/A _____ Date corrected _____

Subpart J—Welding and Cutting

WELDING AND CUTTING—1926.350

1. When transporting or storing compressed gas cylinders, are cylinders secured and caps in place?
   Yes _____ No _____ N/A _____ Date corrected _____

2. Are cylinders secured in a vertical position when transported by power vehicles?
   Yes _____ No _____ N/A _____ Date corrected _____

3. Are all compressed gas cylinders secured in an upright position at all times?
   Yes _____ No _____ N/A _____ Date corrected _____

4. Is it ensured that cylinders, full or empty, are never used as rollers or supports?
   Yes _____ No _____ N/A _____ Date corrected _____

5. Are employers instructed in the safe use of fuel gas?
   Yes _____ No _____ N/A _____ Date corrected _____

6. Are torches inspected for leaking shutoff valves, hose couplings, and tip connections at the beginning of each shift?
   Yes _____ No _____ N/A _____ Date corrected _____

7. Are oxygen cylinders and fittings kept away from oil and grease?
   Yes _____ No _____ N/A _____ Date corrected _____

8. Are oxygen and fuel gas regulators in proper working order?
   Yes _____ No _____ N/A _____ Date corrected _____

9. Are frames of all arc welding and cutting machines grounded?
   Yes _____ No _____ N/A _____ Date corrected _____

10. Are employees instructed in the safe means of arc welding and cutting?
    Yes _____ No _____ N/A _____ Date corrected _____

11. Are welding and cutting operations shielded by noncombustible or flameproof screens whenever practicable?
    Yes _____ No _____ N/A _____ Date corrected _____

12. Are electrodes removed and electrode holders placed or protected so they cannot make electrical contact with employees when the holders are left unattended?
    Yes _____ No _____ N/A _____ Date corrected _____

13. Are employees who are performing any type of welding, cutting, or heating protected by suitable eye protective equipment?
    Yes _____ No _____ N/A _____ Date corrected _____

14. Is suitable fire extinguishing equipment immediately available in work areas and ready for instant use?

Yes _____ No _____ N/A _____ Date corrected _____

15. Are drums, containers, or hollow structures that have contained toxic or flammable substances either filled with water or thoroughly cleaned of such substances, ventilated, and tested before welding, cutting, or heating?

Yes _____ No _____ N/A _____ Date corrected _____

16. Before heat is applied to a drum, container, or hollow structure, is a vent or opening provided to release built-up pressure?

Yes _____ No _____ N/A _____ Date corrected _____

17. Is the mechanical ventilation system of sufficient capacity and so arranged to remove fumes and smoke and to keep the concentration within safe limits?

Yes _____ No _____ N/A _____ Date corrected _____

18. When employees are welding, cutting, or heating in confined space, are general mechanical ventilation, local exhaust ventilation, or airline respirators provided?

Yes _____ No _____ N/A _____ Date corrected _____

Subpart K—Electrical

ELECTRICAL—1926.403

1. Is all electrical equipment free from recognized hazards that might cause death or serious harm?

Yes _____ No _____ N/A _____ Date corrected _____

2. Are disconnecting means legibly marked to indicate purpose unless located so that purpose is evident?

Yes _____ No _____ N/A _____ Date corrected _____

3. Is sufficient working space provided to permit safe operation and maintenance of electrical equipment?

Yes _____ No _____ N/A _____ Date corrected _____

4. Are live electrical parts guarded against accidental contact?

Yes _____ No _____ N/A _____ Date corrected _____

5. Is polarity of conductors correct?

Yes _____ No _____ N/A _____ Date corrected _____

6. Are ground-fault circuit interrupters used to protect employees?

Yes _____ No _____ N/A _____ Date corrected _____

7. If not, is an assured equipment grounding program in place?

Yes _____ No _____ N/A _____ Date corrected _____

8. Are outlet devices correctly matched with load being served?

Yes _____ No _____ N/A _____ Date corrected _____

9. Is path to ground from circuits, equipment, and enclosures permanent and continuous?

Yes _____ No _____ N/A _____ Date corrected _____

10. Are exposed, noncurrent, carrying metal parts of cords and plug-connected equipment rounded?
    Yes _____ No _____ N/A _____ Date corrected _____

11. Are lamps for general illumination protected against breakage?
    Yes _____ No _____ N/A _____ Date corrected _____

12. Are flexible cords and cables protected from damage?
    Yes _____ No _____ N/A _____ Date corrected _____

13. Are electrical extension cords of the three-wire type?
    Yes _____ No _____ N/A _____ Date corrected _____

14. Are unused openings in cabinets, boxes, and fittings closed?
    Yes _____ No _____ N/A _____ Date corrected _____

15. Do all pull boxes, junction boxes, and fittings have covers?
    Yes _____ No _____ N/A _____ Date corrected _____

16. Are all cabinets, cut-out boxes, fittings, boxes, panel board enclosures, switches, circuit breakers, through doorways, or windows attached to building surfaces or concealed behind walls, ceilings, or floors?
    Yes _____ No _____ N/A _____ Date corrected _____

17. Are flexible cords and cables not run through holes in walls, ceilings, and floors, run through doorways or windows, attached to building surfaces, or concealed behind walls, ceilings, or floors?
    Yes _____ No _____ N/A _____ Date corrected _____

18. Are fixtures and receptacles in wet or damp locations identified for that purpose and installed so that water cannot enter?
    Yes _____ No _____ N/A _____ Date corrected _____

19. Is all electrical equipment used in hazardous locations either approved for the location or intrinsically safe?
    Yes _____ No _____ N/A _____ Date corrected _____

20. Are electrical cords or cables taken out of service when worn or frayed?
    Yes _____ No _____ N/A _____ Date corrected _____

Subpart L—Scaffolds

SCAFFOLDS—1926.451

1. Is the footing or anchorage for scaffolds sound, rigid, and capable of supporting the maximum intended load without settling or displacement?
   Yes _____ No _____ N/A _____ Date corrected _____

2. Are scaffold guardrails and toeboards installed on all open sides and ends of platforms more than ten feet above the ground or floor?
   Yes _____ No _____ N/A _____ Date corrected _____

3. Do scaffolds four to ten feet in height, with a minimum horizontal dimension in either direction of less than 45 inches, have standard guardrails on all open sides and ends of platform?
   Yes _____ No _____ N/A _____ Date corrected _____

4. Are scaffolds capable of supporting at least four times their maximum intended load?
    Yes _____ No _____ N/A _____ Date corrected _____

5. Are scaffold planks extended over their end supports not less than six inches, or more than 12 inches?
    Yes _____ No _____ N/A _____ Date corrected _____

6. Are manually propelled mobile scaffolds erected so that their height is no more than four times the minimum base dimension?
    Yes _____ No _____ N/A _____ Date corrected _____

7. Are casters or wheels on mobile scaffolds locked while in use by any person?
    Yes _____ No _____ N/A _____ Date corrected _____

8. Are all two-point suspended scaffolds suspended by wire, synthetic, or fiber ropes capable of supporting at least six times the related load?
    Yes _____ No _____ N/A _____ Date corrected _____

9. Are all ropes (wire, fiber, and synthetic), slings, hangers, platforms, and other supporting parts of two-point suspended scaffolds inspected before every installation?
    Yes _____ No _____ N/A _____ Date corrected _____

10. Are all employees on two-point suspended scaffolds protected by a lifeline and safety belt?
    Yes _____ No _____ N/A _____ Date corrected _____

Subpart M—Fall Protection

FLOOR AND WALL OPENINGS—1926.500

1. Are floor and wall openings properly guarded with standard railings and toe-boards?
    Yes _____ No _____ N/A _____ Date corrected _____

2. Are all holes including skylight openings guarded by fixed standard railings or covers?
    Yes _____ No _____ N/A _____ Date corrected _____

3. Are wall openings four feet or more above ground properly guarded?
    Yes _____ No _____ N/A _____ Date corrected _____

4. Are extension platforms outside a wall properly guarded with side rails or equivalent guards?
    Yes _____ No _____ N/A _____ Date corrected _____

5. Are open-sided floor platforms six feet or more above ground or floor levels guarded by standard railings?
    Yes _____ No _____ N/A _____ Date corrected _____

6. Are runways higher than four feet and, where tools, machine parts, or material are likely to be used, guarded by a standard railing and toeboard?
    Yes _____ No _____ N/A _____ Date corrected _____

7. Are flights of stairs with four or more risers equipped with standard stair railings or handrails?
   Yes _____ No _____ N/A _____ Date corrected _____

8. Are employees performing built-up roofing work on low-pitched roofs with a ground-to-eave height greater than 16 feet protected from falling from the side edge of the roof?
   Yes _____ No _____ N/A _____ Date corrected _____

9. Where mechanical equipment is being used on the roof during built-up roofing work and where a warning line system is used to protect the workers from falling, is the warning line erected around all sides of the work area?
   Yes _____ No _____ N/A _____ Date corrected _____

10. Is the warning line flagged with a high-visibility material at not more than six foot intervals?
    Yes _____ No _____ N/A _____ Date corrected _____

11. Is the warning line supported so that its lowest point (including slack) is no closer than 34 inches or higher than 39 inches from the roof surface?
    Yes _____ No _____ N/A _____ Date corrected _____

12. Are employees working on a roof-edge materials-handling area or materials-storage area protected from falling by guardrails, safety nets, or a safety belt system?
    Yes _____ No _____ N/A _____ Date corrected _____

13. Are materials stored at least six feet away from the edge when guardrails are not erected at the roof edge?
    Yes _____ No _____ N/A _____ Date corrected _____

14. Is a training program designed to train employees in the recognition of hazards of falling that are associated with working near a roof perimeter, and is this program provided for all employees?
    Yes _____ No _____ N/A _____ Date corrected _____

Subpart N—Cranes, Derricks, Hoists, Elevators, and Conveyors

CRANES, DERRICKS, HOISTS, ELEVATORS, AND CONVEYORS—1926.550

1. Are manufacturer's specifications and limitations applicable to the operation of any and all cranes and derricks complied with?
   Yes _____ No _____ N/A _____ Date corrected _____

2. Are rated load capacities, recommended operating speeds, and special hazard warnings posted on all equipment and visible from the operator's station?
   Yes _____ No _____ N/A _____ Date corrected _____

3. Is equipment inspected by a competent person before each use?
   Yes _____ No _____ N/A _____ Date corrected _____

4. Are thorough annual inspections made on hoisting machinery and records of the dates and results of the inspection maintained by employer?
   Yes _____ No _____ N/A _____ Date corrected _____

5. Are accessible areas within the swing radius of the rotating superstructure of the crane barricaded?
Yes _____ No _____ N/A _____ Date corrected _____

6. Are cranes or derricks only used to hoist employees on a personal platform when conventional means are more hazardous or impossible?
Yes _____ No _____ N/A _____ Date corrected _____

7. If a personal platform is being used, are all operation criteria required by this standard being followed?
Yes _____ No _____ N/A _____ Date corrected _____

8. Does the crane or derrick used with a personal platform have a boom angle indicator (if equipped with a variable angle boom), a device to indicate boom length (if equipped with telescoping boom), and an anti-two-blocking device or two-block damage prevention feature?
Yes _____ No _____ N/A _____ Date corrected _____

9. Does the personnel platform meet all design criteria and platform specifications required by this standard?
Yes _____ No _____ N/A _____ Date corrected _____

10. Have a trail lift, an inspection, and proof testing been conducted?
Yes _____ No _____ N/A _____ Date corrected _____

11. Are employees prohibited from riding on material hoists except for the purpose of inspection and maintenance?
Yes _____ No _____ N/A _____ Date corrected _____

12. Are hoistway entrances protected by substantial gates or bars?
Yes _____ No _____ N/A _____ Date corrected _____

13. Are hoistway doors or gates on personnel hoists at least six inches high?
Yes _____ No _____ N/A _____ Date corrected _____

14. Are hoistway doors or gates provided with mechanical locks that cannot be operated from the landing side and are accessible only to persons in the car?
Yes _____ No _____ N/A _____ Date corrected _____

15. Are overhead protective coverings provided on top of hoist cages or platforms?
Yes _____ No _____ N/A _____ Date corrected _____

16. Is the safe working load of overhead hoists, as determined by the manufacturer, indicated on the hoists and not being exceeded?
Yes _____ No _____ N/A _____ Date corrected _____

17. Where conveyors pass over areas or aisles, have guards been provided to protect employees from falling materials?
Yes _____ No _____ N/A _____ Date corrected _____

18. Are conveyors equipped with audible warning signals and is that signal sounded immediately before starting the conveyor?
Yes _____ No _____ N/A _____ Date corrected _____

Subpart O—Motor Vehicles, Mechanized Equipment, and Marine Operations

MOTOR VEHICLES AND MECHANIZED EQUIPMENT—1926.600

1. Are all vehicles that are left unattended at night, adjacent to a highway in normal use or a construction site where work is in progress, equipped with lights, reflectors, or barricades to identify the location of the equipment?
Yes _____ No _____ N/A _____ Date corrected _____

2. Are tire racks, cages, or equivalent protective devices provided and used when inflating, mounting, or dismounting tires installed on split rims or locking rings?
Yes _____ No _____ N/A _____ Date corrected _____

3. Are bulldozer and scraper blades, dump bodies, and so on, fully lowered or blocked when being repaired or not in use?
Yes _____ No _____ N/A _____ Date corrected _____

4. Are parking brakes set on parked equipment, and are wheels chocked when parked on an incline?
Yes _____ No _____ N/A _____ Date corrected _____

5. Are motor vehicles operated within an off-highway job site that is closed to public traffic?
Yes _____ No _____ N/A _____ Date corrected _____

6. Do these vehicles have a service brake system, emergency brake system, and parking brake system in operable condition?
Yes _____ No _____ N/A _____ Date corrected _____

7. Are all vehicles equipped with an audible warning device that is in operable condition at the operator's station?
Yes _____ No _____ N/A _____ Date corrected _____

8. Do all vehicles with an obstructed view to the rear have a back-up alarm, or are they always used with an observer?
Yes _____ No _____ N/A _____ Date corrected _____

9. Do all vehicles have seat belts, and are they used?
Yes _____ No _____ N/A _____ Date corrected _____

10. Are tailgate handles on dump trucks arranged to keep operators clear?
Yes _____ No _____ N/A _____ Date corrected _____

11. Are operating levers on dump trucks equipped with latches?
Yes _____ No _____ N/A _____ Date corrected _____

12. Are vehicles in use inspected at the beginning of each shift to ensure that all parts, equipment, and accessories affecting safety operation are free of defects?
Yes _____ No _____ N/A _____ Date corrected _____

13. Are seat belts provided on all earthmoving equipment except those not equipped with rollover protective structures (ROPS) and those designed for stand-up operation?
Yes _____ No _____ N/A _____ Date corrected _____

14. Does all bidirectional earthmoving equipment have a horn in operable condition?
Yes _____ No _____ N/A _____ Date corrected _____

15. Is all earthmoving or compacting equipment with obstructed rear view equipped with an operable back-up alarm or used only with an observer?
    Yes _____ No _____ N/A _____ Date corrected _____

16. Are all high-lift rider industrial trucks equipped with overhead guards?
    Yes _____ No _____ N/A _____ Date corrected _____

17. Is all equipment used in site-clearing operations equipped with proper rollover protection?
    Yes _____ No _____ N/A _____ Date corrected _____

18. Unless employees can step safely to or from the wharf, float, or river towboat, is a ramp of adequate strength, with side boards, well maintained and properly secured or a safe walkway provided?
    Yes _____ No _____ N/A _____ Date corrected _____

19. Are all powered industrial trucks equipped with inspected and working brakes, steering mechanisms, control mechanisms, warning devices, lights, governors, lift overhead devices, guards, and safety devices?
    Yes _____ No _____ N/A _____ Date corrected _____

Subpart P—Excavations

EXCAVATIONS—1926.651

1. Are all surface encumbrances that might create a hazard removed or supported?
   Yes _____ No _____ N/A _____ Date corrected _____

2. Have all underground utility installations been located?
   Yes _____ No _____ N/A _____ Date corrected _____

3. In trenches more than four feet deep, are stairways, ladders, or ramps located so that travel to them is no more than 25 feet?
   Yes _____ No _____ N/A _____ Date corrected _____

4. Are employees exposed to vehicular traffic wearing warning vests made of reflectorized or high-visibility material?
   Yes _____ No _____ N/A _____ Date corrected _____

5. Is a warning system such as barricades, hand or mechanical signals, or stop logs used when mobile equipment approaches the edge of the excavation?
   Yes _____ No _____ N/A _____ Date corrected _____

6. Are testing and controls used to prevent exposure to hazardous atmospheres?
   Yes _____ No _____ N/A _____ Date corrected _____

7. Are excavation or other materials kept at least two feet from the edge of excavation?
   Yes _____ No _____ N/A _____ Date corrected _____

8. Is excavation inspected daily and after any hazard-increasing occurrence?
   Yes _____ No _____ N/A _____ Date corrected _____

9. Are employees in an excavation five feet deep or more, or with the potential for a cave-in, protected by an adequate protective system?
   Yes _____ No _____ N/A _____ Date corrected _____

Subpart Q—Concrete and Masonry Construction

CONCRETE, CONCRETE FORMS, AND SHORING—1926.701

1. Is all protruding reinforced steel, onto or into which employees could fall, guarded to eliminate the hazard of impalement?
   Yes _____ No _____ N/A _____ Date corrected _____

2. Are employees prohibited from riding concrete buckets?
   Yes _____ No _____ N/A _____ Date corrected _____

3. Do powered, rotating-type concrete trowels, which are manually guided, have a control switch that automatically shuts off if its operator's hands are removed from the handles?
   Yes _____ No _____ N/A _____ Date corrected _____

4. Is a lock-out tag procedure in use for any machinery where inadvertent operation could cause injury?
   Yes _____ No _____ N/A _____ Date corrected _____

5. Is all formwork for cast-in-place concrete designed, fabricated, erected, supported, braced, and maintained so that it will support without failure all loads that might be anticipated?
   Yes _____ No _____ N/A _____ Date corrected _____

6. Is erected shoring equipment inspected immediately prior to, during, and immediately after concrete placement?
   Yes _____ No _____ N/A _____ Date corrected _____

7. Are forms and shores left in place until the employer determines that the concrete can support its weight and superimposed loads?
   Yes _____ No _____ N/A _____ Date corrected _____

8. Are precast concrete wall units, structural framing, and tilt-up wall panels supported to prevent overturning and collapse until permanent connections are made?
   Yes _____ No _____ N/A _____ Date corrected _____

9. Do designs and plans include prescribed methods of erection?
   Yes _____ No _____ N/A _____ Date corrected _____

10. Does jacking equipment have a safety factor of 2.5?
    Yes _____ No _____ N/A _____ Date corrected _____

11. Is the maximum number of manually controlled jacks on one slab limited to 14?
    Yes _____ No _____ N/A _____ Date corrected _____

12. Are jacking operations synchronized to ensure even and uniform lifting?
    Yes _____ No _____ N/A _____ Date corrected _____

13. Are only those employees required for jacking and to secure slabs permitted under the slab during jacking?
    Yes _____ No _____ N/A _____ Date corrected _____

14. Is a limited access zone established when constructing a masonry wall?
    Yes _____ No _____ N/A _____ Date corrected _____

15. Are all masonry walls over eight feet in height braced or supported to prevent collapse?
    Yes _____ No _____ N/A _____ Date corrected _____

Subpart R—Steel Erection

STEEL ERECTION—1926.750

1. Are safety nets used when the work area is more than 25 feet above the ground, water surface, or other surface where ladders, scaffolds, catch platforms, temporary floors, safety lines, and safety belts are impractical?
   Yes _____ No _____ N/A _____ Date corrected _____

2. Is permanent flooring installed as the erection progresses, and is there a maximum of eight floors between the erection floor and the uppermost permanent floor, except where the structural integrity is maintained as a result of the design?
   Yes _____ No _____ N/A _____ Date corrected _____

3. Is the derrick or erection floor solidly planked except for access openings?
   Yes _____ No _____ N/A _____ Date corrected _____

4. Is the planking or decking of proper thickness to carry the work load?
   Yes _____ No _____ N/A _____ Date corrected _____

5. Is planking two inches minimum, full-size undressed, laid tight, and secured?
   Yes _____ No _____ N/A _____ Date corrected _____

6. Is a safety railing of one-half-inch wire rope or equal installed approximately 42 inches around the periphery of all temporary planked or temporary metal-decked floors of tiered buildings and other multifloored structures during structural steel assembly?
   Yes _____ No _____ N/A _____ Date corrected _____

7. Where long span joints or trusses 40 feet or longer are used, is a center row of bolted bridging installed?
   Yes _____ No _____ N/A _____ Date corrected _____

8. Are tag lines used for controlling loads?
   Yes _____ No _____ N/A _____ Date corrected _____

9. Are locking devices provided to retain sockets on impact wrenches?
   Yes _____ No _____ N/A _____ Date corrected _____

10. When riveting in the vicinity of combustible material, are precautions taken to prevent fires?
    Yes _____ No _____ N/A _____ Date corrected _____

11. On pneumatic riveting hammers, is the safety wire on snap not less than number 14 wire, and on handle not less than number 9 wire?
    Yes _____ No _____ N/A _____ Date corrected _____

12. Are turnbuckles secured to prevent unwinding under stress?
    Yes _____ No _____ N/A _____ Date corrected _____

13. Are plumbing-up guys and related equipment placed so that employees can reach connection points?
    Yes _____ No _____ N/A _____ Date corrected _____

14. In plumbing up, do the planks overlap the bearing on each end by a minimum of 12 inches?
    Yes _____ No _____ N/A _____ Date corrected _____

13. Is wire mesh, exterior plywood, or the equivalent placed around columns where planks do not fit tightly?
Yes _____ No _____ N/A _____ Date corrected _____

14. Are all unused openings in floor planked over or guarded?
Yes _____ No _____ N/A _____ Date corrected _____

15. Are employees provided with safety belts when working on float scaffolds?
Yes _____ No _____ N/A _____ Date corrected _____

Subpart S—Underground Construction, Caissons, Cofferdams, and Compressed Air

UNDERGROUND CONSTRUCTION—1926.800

1. Are safe means of access provided and maintained for all working places?
Yes _____ No _____ N/A _____ Date corrected _____

2. Is a check-in and check-out system used that will provide positive identification of every employee underground? Is an accurate record and location of the employees kept on the surface?
Yes _____ No _____ N/A _____ Date corrected _____

3. Are emergency evacuation plans and procedures developed and made known to employees?
Yes _____ No _____ N/A _____ Date corrected _____

4. Are Bureau of Mines–approved self-rescuers available to equip each employee near the advancing face and on haulage equipment and other areas where employees might be trapped by smoke or gas?
Yes _____ No _____ N/A _____ Date corrected _____

5. Is a maximum of one day's supply of diesel fuel stored underground?
Yes _____ No _____ N/A _____ Date corrected _____

6. Are gasoline and liquefied petroleum gases prohibited from being taken, stored, or used underground?
Yes _____ No _____ N/A _____ Date corrected _____

7. Are enclosed metal cages used to raise and lower persons in the shaft?
Yes _____ No _____ N/A _____ Date corrected _____

8. At cofferdams, are warning signals for evacuation of employees in case of emergency developed and posted?
Yes _____ No _____ N/A _____ Date corrected _____

9. Is a competent person present at all times who is designated and represents the employer, and who is familiar with all requirements of compressed air?
Yes _____ No _____ N/A _____ Date corrected _____

Subpart T—Demolition

DEMOLITION—1926.850

1. If employees are exposed to the hazard of falling through wall openings, are the openings protected to a height of approximately 42 inches?
Yes _____ No _____ N/A _____ Date corrected _____

2. If debris is dropped through holes in the floor without chutes, is the area onto which the material is dropped completely enclosed with barricades at least 42 inches high and at least six feet back from the projected edge of the opening above?
Yes _____ No _____ N/A _____ Date corrected _____

3. Are all floor openings not used as material drops equipped with a properly secured cover that will support any load that might be imposed?
Yes _____ No _____ N/A _____ Date corrected _____

4. Are all stairs, passageways, ladders, and incidental equipment covered by this section periodically inspected and maintained in a clean, safe condition?
Yes _____ No _____ N/A _____ Date corrected _____

5. Is any area where material is dropped outside the exterior walls of the structure effectively protected?
Yes _____ No _____ N/A _____ Date corrected _____

6. Are workers engaged in razing the steel after floor arches are removed protected by planking?
Yes _____ No _____ N/A _____ Date corrected _____

7. Are continuous inspections made by a competent person as work progresses to detect hazards from weakened or deteriorated floors or walls or loosened materials?
Yes _____ No _____ N/A _____ Date corrected _____

Subpart U—Blasting and the Use of Explosives

BLASTING AND EXPLOSIVES—1926.900

1. Are only authorized and qualified persons permitted to handle explosives?
Yes _____ No _____ N/A _____ Date corrected _____

2. Are smoking, firearms, matches, open-flame lamps and other fires, flame- or heat-producing devices, and sparks prohibited in or near explosive magazines and while explosives are being handled, transported, or used?
Yes _____ No _____ N/A _____ Date corrected _____

3. Is an inventory and use record of all explosives maintained by the employer?
Yes _____ No _____ N/A _____ Date corrected _____

4. Are explosives not in use kept in a locked magazine?
Yes _____ No _____ N/A _____ Date corrected _____

5. Are precautions taken to prevent accidental discharge of electric blasting caps from current induced by radar, radio transmitters, lighting, adjacent power lines, dust storms, and other sources of extraneous electricity?
    Yes _____ No _____ N/A _____ Date corrected _____

6. Do all blasters meet the requirements specified by this standard?
    Yes _____ No _____ N/A _____ Date corrected _____

7. Is every vehicle or conveyance used for transporting explosives marked on both sides, front, and rear with placards reading "Explosives" in red letters not less than four inches high on white background?
    Yes _____ No _____ N/A _____ Date corrected _____

8. Are motor vehicles transporting explosives always attended?
    Yes _____ No _____ N/A _____ Date corrected _____

9. Are explosives and related materials stored in approved facilities?
    Yes _____ No _____ N/A _____ Date corrected _____

10. Are blasting caps, electric blasting caps, detonating primers, and primed cartridges stored in separate magazines from explosives or blasting agent?
    Yes _____ No _____ N/A _____ Date corrected _____

11. Is tamping done only with wood rods or plastic tamping poles without exposed metal parts except for nonsparking metal connections of jointed poles?
    Yes _____ No _____ N/A _____ Date corrected _____

12. Is the so-called "drop fuse" method of dropping or pushing primer or any explosive with a lighted fuse prohibited?
    Yes _____ No _____ N/A _____ Date corrected _____

13. Is a loud warning signal given by the blaster in charge before that blast is fired?
    Yes _____ No _____ N/A _____ Date corrected _____

Subpart V—Power Transmission and Distribution

POWER TRANSMISSION AND DISTRIBUTION—1926.950

1. Are electric equipment and lines considered energized until determined to be de-energized by test or other appropriate methods or means?
    Yes _____ No _____ N/A _____ Date corrected _____

2. Does the employer provide training or require that his or her employees are knowledgeable and proficient in procedures involving emergency situations and first aid fundamentals?
    Yes _____ No _____ N/A _____ Date corrected _____

3. Are aerial lift trucks working near energized lines or equipment grounded or barricaded and considered as energized equipment, or are the trucks insulated for the work being performed?
    Yes _____ No _____ N/A _____ Date corrected _____

4. Are tag lines or other suitable devices used to control loads being handled by hoisting equipment where hazards to employees exist?
    Yes _____ No _____ N/A _____ Date corrected _____

5. When attaching grounds, is the ground end attached first and the end attached and removed using insulated tools or other suitable devices?
   Yes _____ No _____ N/A _____ Date corrected _____
6. When working on buried cable or a cable in manholes, is metallic sheath continuity maintained by bonding across the opening or by equivalent means?
   Yes _____ No _____ N/A _____ Date corrected _____
7. Are the requirements of paragraphs (a) and (b) of this section complied with for all lineman body belts, safety straps, and lanyards?
   Yes _____ No _____ N/A _____ Date corrected _____

Subpart W—Rollover Protective Structures; Overhead Protection

ROLLOVER PROTECTIVE STRUCTURES (ROPS)—1926.1001

1. Are all rubber-tired, self-propelled scrapers, rubber-tired front-end loaders, wheel-type agricultural and industrial tractors, crawler tractors, crawler-type loaders, and motor graders (with or without attachment) equipped with rollover protective structures?
   Yes _____ No _____ N/A _____ Date corrected _____
2. Do ROPS meet minimum performance criteria detailed in these standards?
   Yes _____ No _____ N/A _____ Date corrected _____

CHAPTER 7

Sample Construction Safety Program

It is necessary that your company establish a formal, written safety and health program. It has been well established that an effective safety and health program is the best way to reduce workplace injuries. The elements of an effective program should include the following:

1. Management leadership and employee involvement: Assign safety and health responsibility and authority to supervisors and employees and hold them accountable.
2. Worksite analysis: Identifies current and potential hazards. It includes a thorough baseline survey to identify unsafe acts and conditions, job hazard analysis (written safe operating procedures for major tasks), a self-inspection program, a system for reporting hazards, accident and incident investigation, and analysis of injuries and illnesses.

3. Hazard prevention and control: Prevention consists of regular maintenance and housekeeping, emergency planning and preparation, first aid, and ready access to emergency care. Control includes machine guards, enclosures, locks, protective equipment, safe work procedures, and placement of personnel to minimize hazards.

4. Training: Training of all personnel, from managers through supervisors to employees, about the hazards they might be exposed to, and their identification, prevention, and control. Managers and supervisors also need training in program management (e.g., enforcing rules, conducting drills).

> This sample program is provided to give assistance in developing a written construction safety program. Because all construction firms differ in many aspects, each contractor should tailor their own program and formulate safety procedures and rules applicable to their own conditions and work environments.

This is only a sample and should not be used as is. Failure to develop a written construction safety program specific to your operation may result in an Occupational Safety and Health Administration (OSHA) violation.

It must be noted that the final determination of compliance with OSHA regulations, including the written safety and health program, is made by evaluation of all factors pertaining to a particular worksite with respect to employee safety and health. Employers who use this guide should be aware that it is not to be considered a substitute for any provisions of the Occupational Safety and Health Act or for any standards issued by OSHA.

Subcontractor Compliance

All contracts and subcontracts require that state and federal laws concerning health and safety will be observed by the subcontractor. The provisions of these health and safety responsibilities apply to subcontractors and their employees working for this company. Failure to fulfill this requirement is a failure to meet the conditions of the contract.

CONSTRUCTION SAFETY PROGRAM FOR

[company name]

Emergency Response Plan

> This must be filled out *before* beginning work on each site.

For _____ job site

City/location: _____

Subdivision: _____

Street name: _____

Job address: _____

Job phone contact: _____

EMERGENCY PHONE CONTACT NUMBERS

Local fire department/EMS area: _____

Ambulance service: _____

Nearest medical treatment: _____

Directions (EMS/clinic/Dr.): _____

_____

_____

_____

Directions to worksite: _____

_____

_____

_____

EMERGENCY RESPONSE TO HAZARDOUS SUBSTANCES

> For small construction companies not normally involved in hazardous waste cleanups.

If any substance is found of unknown origin, company policy is to *leave it alone*! Immediately evacuate the area, and contact the nearest hazardous material response team. Do not allow employees on site until declared safe by the response team.

First Aid

- Arrangements must be made *before* starting the project, to provide for prompt medical response in the event of an emergency.

- In areas where severe bleeding, suffocation, or severe electrical shock can occur, a three- to four-minute response time is required.
- If medical attention is not available within four minutes, then a first aid–trained person must be available on the job site at all times.
- An appropriate, weatherproof first aid kit must be on site. It must be checked weekly.
- Provisions for an ambulance or other transportation must be made in advance.
- Contact methods must be provided.
- Telephone numbers must be posted where 911 is not available.

[Company name] has designated [safety person or foreman] as having adequate training to render first aid in the event of a medical emergency in areas where emergency response time is in excess of four minutes. They will maintain appropriate first aid kits and check them weekly to ensure they are properly stocked.

First aid kits are located at the following locations:

- _____
- _____
- _____

Every employee shall be trained in emergency procedures:

- Evacuation plan
- Alarm systems
- Shutdown procedures for equipment
- Types of potential emergencies

It is the employer's responsibility to review the job sites, addressing all potential emergency situations.

Policy Statement

It is [company name]'s belief that our people are our most important asset and that the preservation of employee safety and health must remain a constant consideration in every phase of our business. We will provide the resources necessary to manage, control, or eliminate all safety and health hazards.

All employees are responsible for working safely and productively, as well as recognizing and being aware of hazards in their work areas. Employees are also responsible for following safe work practices, including the use of personal protective equipment (PPE) where necessary.

It is our belief that any safety and health program must have total employee involvement. Therefore, this program has management's highest priority, support, and participation.

Production is not so urgent that we cannot take time to do our work safely.

[Company name] President

Goals

Safety begins at the top and goes downward throughout the company. The primary goal of [company name] is to continue operating a profitable business while protecting employees from injuries, illness, or harm. This can be achieved in part by delegating responsibility and accountability to all involved in this company's operation.

- Responsibility: Having to answer for activities and results.
- Accountability: The actions taken by management to ensure the performance of responsibilities.

In other words, to reach our goal of a safe workplace, everyone needs to take responsibility and be held accountable.

Benefits of achieving our goals are as follows:

- Minimizing injuries and accidents;
- Minimizing the loss of property and equipment;
- Elimination of potential fatalities;
- Elimination of potential permanent disabilities;
- Elimination of potential OSHA fines;
- Reductions in workers' compensation costs;
- Reductions in operating costs; and
- Having the best safety and health conditions possible in the workplace.

Management Commitment

The management of [company name] is committed to the company's safety policy and to provide direction and motivation by

- appointing [safety person] as our safety coordinator;
- establishing company safety goals and objectives;
- developing and implementing a written safety and health program;
- ensuring total commitment to the safety and health program;
- facilitating employees' safety training;
- establishing responsibilities for management and employees to follow;
- ensuring that management and employees are held accountable for performance of their safety responsibilities;
- establishing and enforcing disciplinary procedures for employees; and
- reviewing the safety and health program annually and revising or updating as needed.

Safety Committee and Safety Meetings

The committee shall consist of representatives from management and nonmanagement employees with [safety person] as the chairman. The committee is a forum, created for the purpose of fostering safety and health through communication.

The responsibilities of safety committee members include the following:

- Discussing safety policies and procedures with management and making recommendations for improvements;
- Reviewing accident investigation reports on all accidents and "near misses"; and
- Identifying unsafe conditions and work practices and making recommendations for corrections.

All employees of [company name] shall attend and participate in the weekly safety meetings. The weekly safety meeting shall be conducted by [safety person]. Problems that have arisen or that are anticipated shall be discussed along with any other safety and health topics. The meeting shall be kept a valuable educational experience by

- keeping the meetings moving;
- starting and stopping on time;
- using illustrated material and demonstrations to make the point;
- discussing each topic thoroughly, providing handouts if possible;
- reviewing accidents, injuries, property losses, and "near misses"; and
- evaluating accidents, injuries, property losses, and "near misses" for trends and similar causes to initiate corrective actions.

The safety coordinator must document the meetings using the form in appendix A.

Assignment of Responsibility

[Company name] has designated [safety person] as our safety coordinator. Their cell phone and office phone numbers are as follows:

Office: (xxx) xxx-xxxx
Cell: (xxx) xxx-xxxx

It shall be the duty of the safety coordinator to assist the supervisor or foreman and all other levels of management in the initiation, education, and execution of an effective safety program including the following:

- Introducing the safety program to new employees;
- Following up on recommendations, suggestions, and so forth, made at the weekly safety meetings (all topics of safety concerns must be documented accordingly);
- Assisting the personnel in the execution of standard policies;
- Conducting safety inspections on a periodic basis;
- Addressing all hazards or potential hazards as needed;
- Preparing monthly accident reports and investigations;

- Maintaining adequate stock of first aid supplies and other safety equipment to ensure their immediate availability;
- Making sure there is an adequate number of qualified first aid–certified people on the worksite;
- Becoming thoroughly familiar with OSHA regulations and local and state safety codes;
- Defining the responsibilities for safety and health of all subordinates and holding each person accountable for their results through the formal appraisal system and, where necessary, disciplinary procedures; and
- Emphasizing to employees that accidents create unnecessary personal and financial losses.

SUPERVISORS AND FOREMEN

The supervisors and foremen will establish an operating atmosphere that ensures that safety and health is managed in the same manner and with the same emphasis as production, cost, and quality control.

- Regularly emphasize that accident and health hazard exposure prevention are not only moral responsibilities but also a condition of employment.
- Identify operational oversights that could contribute to accidents that often result in injuries and property damage.
- Participate in safety and health related activities, including routinely attending safety meetings, reviews of the facility, and correcting employee behavior that can result in accidents and injuries.
- Spend time with each person hired explaining the safety policies and the hazards of his or her particular work.
- Ensure that initial orientation of "new hires" is carried out by [safety person].
- Make sure that, if a "competent person" is required, one is present to oversee and instruct employees when necessary.
- Never shortcut safety for expediency or allow workers to do so.
- Enforce safety rules consistently and follow the company's discipline and enforcement procedures.
- Conduct a daily, job-site safety inspection and correct noted safety violations.

EMPLOYEES

It is the duty of each and every employee to know the safety rules and to conduct his or her work in compliance with these rules. Disregard of the safety and health rules shall be grounds for disciplinary action up to and including termination. It is also the duty of each employee to make full use of the safeguards provided for their protection. Every employee will receive an orientation when hired and receive a copy of the company safety and health program. Employee responsibilities include the following:

- Reading, understanding, and following safety and health rules and procedures;
- Signing the policies and procedures acknowledgment included in appendix B;
- Wearing personal protective equipment (PPE) at all times when working in areas where there is a possible danger of injury;
- Wearing suitable work clothes as determined by the supervisor or foreman;
- Performing all tasks safely as directed by the supervisor or foreman;
- Reporting *all* injuries, no matter how slight, to the supervisor or foreman immediately and seeking treatment promptly;
- Knowing the location of first aid, fire-fighting equipment, and other safety devices;
- Attending any and all required safety and health meetings; and
- Not performing potentially hazardous tasks, or using any hazardous material until properly trained, and following all safety procedures when performing those tasks.

Discipline and Enforcement

[Company name] seeks to establish and maintain standards of employee conduct and supervisory practices that will support and promote safe and effective business operations. These supervisory practices include administering corrective action when employee safety performance or conduct jeopardizes this goal. This policy sets forth general guidelines for a corrective action process aimed to document and correct undesirable employee behavior. Major elements of this policy include the following:

- Constructive criticism and instruction by the employee's supervisor or foreman to educate and inform employees of appropriate safety performance and behavior;
- Correcting employee's negative behavior to the extent required;
- Informing the employee that continued violation of company safety policies might result in termination; and
- Written documentation of disciplinary warnings and corrective action taken.

Depending on the facts and circumstances involved with each situation, the company might choose any corrective action including immediate termination. However, in most circumstances, the subsequent steps will be followed:

1. Verbal warning, informally documented, by a supervisor, foreman, or safety coordinator for minor infractions of company safety rules. A supervisor, foreman, or safety coordinator must inform the employee what safety rule or policy was violated and how to correct the problem.
2. Written warning, documented in the employee's file. Repeated minor infractions or a more substantial safety infraction requires issuance of a written warning. Every attempt should be made to reeducate the employee on the desired performance. The employee should acknowledge the warning by signing the document before it is placed in his or her personnel file.

3. Suspension, for three working days, if the employee fails to appropriately respond or management determines the infraction is sufficiently serious.
4. Termination, for repeated or serious safety infractions.

Control of Hazards

Where feasible, workplace hazards are prevented by effective design of the job site or job. Where it is not feasible to eliminate such hazards, they must be controlled to prevent unsafe and unhealthy exposure. Once a potential hazard is recognized, the elimination or control must be done in a timely manner. These procedures include measures such as the following:

- Maintaining all extension cords and equipment;
- Ensuring all guards and safety devices are working;
- Periodically inspecting the worksite for safety hazards;
- Establishing a medical program that provides applicable first aid to the site, as well as nearby physician and emergency phone numbers; and
- Addressing any and all safety hazards with employees.

Fire Prevention

Fire prevention is an important part of protecting employees and company assets. Fire hazards must be controlled to prevent unsafe conditions. Once a potential hazard is recognized, it must be eliminated or controlled in a timely manner. The following fire prevention requirements must be met for each site:

- One conspicuously located 2A fire extinguisher (or equivalent) for every floor;
- One 2A conspicuously located fire extinguisher (or equivalent) for every 3,000 square feet;
- A conspicuously located, 10B fire extinguisher for areas in which more than five gallons of flammable liquids or gas are stored;
- Generators and internal combustion engines located away from combustible materials;
- Site free from accumulation of combustible materials or weeds;
- No obstructions or combustible materials piled in the exits;
- No more than 25 gallons of combustible liquids stored on site;
- No liquefied petroleum gas (LP-Gas) containers stored in any buildings or enclosed spaces;
- Fire extinguishers in the immediate vicinity where welding, cutting, or heating is being done;
- Fire-fighting equipment conspicuously located, accessible, and inspected periodically, and maintained in operating condition (an annual service check and monthly visual inspections are required for fire extinguishers);

- All employees knowing the location of fire-fighting equipment in the work area and having knowledge of its use and application;
- Only approved safety cans used for handling or storing flammable liquids in quantities greater than one gallon (for one or less gallon, only the original container or a safety can will be used);
- When heat-producing equipment is used, work areas kept clear of all fire hazards, and all sources of potential fires eliminated;
- A salamander or other open-flame device not used in confined or enclosed structures without proper ventilation (heaters will be vented to the atmosphere and located an adequate distance from walls, ceilings, and floors);
- Fire extinguishers available at all times when utilizing heat-producing equipment; and
- Storage of LP-Gas within buildings prohibited.

Training and Education

Training is an essential component of an effective safety and health program, addressing the responsibilities of both management and employees at the site. Training is most effective when incorporated into other education on performance requirements and job practices.

Training programs should be provided as follows:

- Initially when the safety and health plan is developed;
- For all new employees before beginning work;
- When new equipment, materials, or processes are introduced;
- When procedures have been updated or revised;
- When experiences or operations show that employee performance must be improved; and
- At least annually.

Besides the standard training, employees should also be trained in the recognition of hazards—to be able to look at an operation and identify unsafe acts and conditions. A list of typical hazards employees should be able to recognize might include the following:

- Fall hazards—falls from floors, roofs and roof openings, ladders (straight and step), scaffolds, wall openings, tripping, trenches, steel erection, stairs, chairs
- Electrical hazards—appliances, damaged cords, outlets, overloads, overhead high voltage, extension cords, portable tools (broken casing or damaged wiring), grounding, metal boxes, switches, ground-fault circuit interrupters (GFCI)
- Housekeeping issues—exits, walkways, floors, trash, storage of materials (hazardous and nonhazardous), protruding nails
- Fire hazards—oily or dirty rags, combustibles, fuel gas cylinders, exits (blocked), trips or slips stairs, uneven flooring, electrical cords, icy walkways

- Health hazards—silicosis, asbestos, loss of hearing, eye injury due to flying objects

Employees trained in the recognition and reporting of hazards and supervisors and foremen trained in the correction of hazards will substantially reduce the likelihood of a serious injury.

Recordkeeping and OSHA Log Review

In the event of a fatality (death on the job) or catastrophe (accident resulting in hospitalization of three or more workers) contact [safety person]. Their office and cell-phone numbers are as follows:

Office: (xxx) xxx-xxxx
Cell: (xxx) xxx-xxxx

The safety coordinator will in turn report it to the OSHA Region Office within eight hours after the occurrence.

If an injury or accident should ever occur, you are to report it to your supervisor or foreman as soon as possible. A log entry and summary report shall be maintained for every recordable injury and illness. The entry should be done within seven days after the injury or illness has occurred. The OSHA 300 form or equivalent shall be used for the recording.

An OSHA recordable injury or illness is defined as an injury resulting in loss of consciousness, days away from work, days of restricted work, or medical treatment beyond first aid. First aid includes

- tetanus shots;
- Band-Aids or butterfly bandages;
- cleaning, flushing, or soaking wounds;
- ace bandages and wraps;
- nonprescription drugs at nonprescription strength (Aspirin, Tylenol, etc.);
- drilling fingernails or toenails;
- eye patches, eye flushing, and foreign body removal from eye with Q-tips;
- finger guards;
- hot or cold packs; and
- drinking fluids for heat stress.

An annual summary of recordable injuries and illnesses must be posted at a conspicuous location in the workplace and contain the following information: calendar year, company name or establishment name, establishment address, certifying signature, title, and date. If no injury or illness occurred in the year, zeroes must be entered on the total line.

The OSHA logs should be evaluated by the employer to determine trends or patterns in injuries in order to appropriately address hazards and implement prevention strategies.

Accident Investigation

SUPERVISORS AND FOREMEN

- Provide first aid; call for emergency medical care if required.
- If further medical treatment is required, arrange to have an employer representative accompany the injured employee to the medical facility.
- Secure area, equipment, and personnel from injury and further damage.
- Contact safety coordinator.

SAFETY COORDINATOR

- Investigate the incident (injury)—gather facts and employee and witness statements; take pictures and physical measurements of incident site and equipment involved.
- Complete an incident investigation report form (included in appendix C) and the necessary workers' compensation paperwork within 24 hours whenever possible.
- Ensure that corrective action to prevent a recurrence is taken.
- Discuss incident, where appropriate, in safety and other employee meetings with the intent to prevent a recurrence.
- Discuss incident with other supervisors or foremen and other management.
- If the injury warrants time away from work, ensure that the absence is authorized by a physician and that you maintain contact with your employee while he or she remains off work.
- Monitor status of employees off work, maintain contact with employees, and encourage their return to work even if restrictions are imposed by the physician.
- When injured employees return to work they should not be allowed to return to work without "return to work" release forms from the physician. Review the release carefully and ensure that you can accommodate the restrictions and that the employee follows the restrictions indicated by the physician.

Workers' Compensation Claims Management

The following actions will be followed on all accidents or injuries being submitted as a workers' compensation claim.

- Injured employees must report all accidents and injuries to their supervisor immediately (within 72 hours), who in turn will notify other appropriate company officials, such as the safety manager or claims manager. All accidents and incidents will be investigated by the safety manager, supervisor, or the claims manager to determine the facts and take corrective action to prevent recurrence.
- Employees, within ten days after notification to the employer, must complete the Worker Information section only of the Workers' Safety and Compensation Report of Occupational Injury or Disease forms.

- The supervisor or claims manager will complete the Employer's Information section of the same report within ten days of the notification.
- The claims manager will ensure that management is notified as appropriate by filing the above report within ten days of the notification.
- The accident investigation must confirm that the injury was job related for the resultant claim to be valid.
- Injured employees will be entered into a modified job program, that is, light duty, restricted duty, part-time duty, when such is recommended by the attending physician.

Safety Rules and Procedures

- No employee is expected to undertake a job until that person has received adequate training.
- All employees shall be trained on every potential hazard that they could be exposed to and how to protect themselves.
- No employee is required to work under conditions that are unsanitary, dangerous, or hazardous to health.
- Only qualified trained personnel are permitted to operate machinery or equipment.
- All injuries must be reported to your supervisor or foreman.
- Manufacturer's specifications, limitations, and instructions shall be followed.
- Particular attention should be given to new employees and to employees moving to new jobs or doing nonroutine tasks.
- All OSHA posters shall be posted.
- Emergency numbers shall be posted and reviewed with employees.
- Each employee in an excavation or trench shall be protected from cave-ins by an adequate protective system.
- Employees working in areas where there is a possible danger of head injury, excessive noise exposure, or potential eye and face injury shall be protected by personal protection equipment (PPE).
- All hand and power tools and similar equipment, whether furnished by the employer or the employee, shall be maintained in a safe condition.
- All materials stored in tiers shall be stacked, racked, blocked, interlocked, or otherwise secured to prevent sliding, falling, or collapse.
- The employer shall ensure that electrical equipment is free from recognized hazards that are likely to cause death or serious physical harm to employees.
- All scaffolding shall be erected in accordance with CFR 1926.451 Subpart L. Standard guardrails for fall protection and ladders for safe access shall be used.
- All places of employment shall be kept clean; the floor of every workroom shall be maintained, so far as practicable, in a dry condition; and standing water shall be removed. Where wet processes are used, drainage shall be maintained, and false floors, platforms, mats, or other dry standing places or appropriate waterproof footgear shall be provided.

- To facilitate cleaning, every floor, working place, and passageway shall be kept free from protruding nails, splinters, loose boards, and holes and openings.
- All floor openings, open-sided floor and wall openings, shall be guarded by a standard railings and toeboards or cover.
- The employer shall comply with the manufacturer's specifications and limitations applicable to the operation of any and all cranes and derricks.
- All equipment left unattended at night, adjacent to a highway in normal use, or adjacent to construction areas where work is in progress shall have appropriate lights or reflectors, or barricades equipped with appropriate lights or reflectors, to identify the location of the equipment.
- No construction loads shall be placed on a concrete structure or portion of a concrete structure unless the employer determines, based on information received from a person who is qualified in structural design, that the structure or portion of the structure is capable of supporting the loads.
- A stairway or ladder shall be provided at all personnel points of access where there is a break in elevation of 19 inches or more, and no ramp, runway, sloped embankment, or personnel hoist is provided.

Employee Emergency Action Plan for Fire and Other Emergencies

> The following emergency action plan is appropriate only for small construction sites; larger sites should have a much more detailed plan.

1. Emergency escape procedures: Immediately leave the building through the closest practical exit. Meet up at the foremen's truck.
2. Critical plant operations: Shut off the generator on your way out if possible, otherwise evacuate the building.
3. Accounting for employees: The foreman or supervisor is to account for all employees after emergency evacuation has been completed and assign duties as necessary.
4. Means of reporting fires and other emergencies: Dial 911 on a cell phone, report the location of the emergency, and provide directions to the responders.
5. Further information: Contact the safety coordinator for further information or explanation of duties under the plan.

ALARMS SYSTEMS AND EVACUATION

[Company name] establishes the call: fire, fire, fire (el fuego, el fuego, el fuego) by any employee, as the signal to immediately evacuate the building or facility for fire and other emergencies.

TRAINING

Before implementing the emergency action plan, a sufficient number of persons to assist in the safe and orderly emergency evacuation of employees will be designated and trained.

The plan will be reviewed with each employee covered by the plan at the following times:

- Initially when the plan is developed or upon initial assignment;
- Whenever the employee's responsibilities or designated actions under the plan change; and
- Whenever the plan is changed.

The plan will be kept at the worksite and made available for employee review.

> For those employers with ten or fewer employees the emergency action plan may be communicated orally to employees, and the employer need not maintain a written plan.

Drug-Free Workplace

The following is the company's drug-free workplace policy. Employers are not required to pay the costs of treatment or any other intervention program.

- The unlawful use, possession, transfer, or sale of illegal drugs or controlled substances and the misuse of alcohol by employees during work hours are prohibited.
- The consequences for violation of the drug-free policy might include, but are not limited to, a referral for therapeutic help, discipline, or discharge.
- A list of community resources that provide substance abuse treatment and prevention services is posted at the bulletin board where they may be regularly viewed by employees.
- Encourage the designation of a totally or partially smoke-free workplace.

Driving Safety

Vehicle operations are an integral part of our business. Therefore, the following rules shall apply to all business vehicle operations. Hopefully, employees will follow these rules when operating their own personal vehicles.

- All vehicle operators are required to have a current and valid drivers' license for the vehicle to be operated, that is, motorcycle drivers' license, truck drivers' license, or commercial drivers' license (CDL).

- No unauthorized use of company vehicles shall be permitted.
- All cargo or other items, that is, laptops, suitcases, and so forth, shall be loaded and secured to prevent them from creating hazards in the event of hard braking.
- Prior to entering the vehicle, visually inspect the entire vehicle. Look for broken windows, light covers, low tire pressure, and so on. Report all damage to your supervisor.
- Adjust all mirrors for the proper vision of the operator.
- All occupants shall fasten their seat belts. The vehicle shall not be started until all occupants have fastened their seat belts.
- Check all gauges and switches for proper function and location, that is, cruise control, windshield wipers, lights, gearshift, and radio. Do not look for these while you are operating the vehicle. Test the brakes to determine their effectiveness and get a "feel" for the necessary brake pressure.
- Obey all traffic laws while operating the vehicle. This includes the speed limit.
- Vehicles shall *not* be operated while under the influence of alcohol or drugs that may impair your driving ability. Some prescription drugs and over-the-counter drugs might also affect your driving and decision-making abilities.
- Cell-phone operation must be conducted *only* while stopped and out of traffic.
- Pay attention! Keep your mind on driving and watching the road. Watch out for other drivers. Make sure you are well rested and alert.
- Don't get involved in "road rage." Don't become angry at aggressive drivers. Simply pull over to the right lane or the side of the road and allow them to pass.
- Always stay at least two seconds behind the vehicle in front of you. If driving conditions are not optimal, that is, rain, ice, snow, wind, or low visibility, allow a further following distance.
- If your vehicle becomes disabled, call for help on your cell phone or display a white flag on the antenna as a request for help. Require identification of strangers who offer assistance.
- Keep your doors locked. Park in well-lighted areas. Have your keys ready to enter your vehicle. You are a target when looking in your purse or digging in a handbag.
- When approaching your vehicle, try to observe any persons in the vicinity of your vehicle and look under your vehicle. Look in the back seat before opening the door. Carry a penlight flashlight.
- Vary your routes and schedules.
- Leave an itinerary of your trip with your supervisor or family member.

Construction Safety and Health Rules

In order for a safety program to be effective, it is vital that it be understood and implemented at all levels from management to all employees.

The following are the occupational safety and health rules and regulations applicable to our operations that must be complied with by our company. A complete set of safety standards can be found in OSHA's Regulations for Construction (1926) Industry.

GENERAL WORKPLACE SAFETY RULES

- Report unsafe conditions to your immediate supervisor.
- Promptly report all accidents, injuries, and incidents to your immediate supervisor.
- Use eye and face protection where there is danger from flying objects or particles (such as when grinding, chipping, burning and welding, and so forth) or from hazardous chemical splashes.
- Dress properly. Wear appropriate work clothes, gloves, and shoes or boots. Loose clothing and jewelry shall not be worn.
- Operate machines or other equipment only when all guards and safety devices are in place and in proper operating condition.
- Keep all equipment in safe working condition. Never use defective tools or equipment. Report any defective tools or equipment to your immediate supervisor.
- Properly care for and be responsible for all personal protective equipment (PPE). Wear or use any such PPE when required.
- Lockout, tagout, or disconnect power on any equipment or machines before any maintenance, unjamming, and adjustments are made.
- Do not leave materials in aisles, walkways, stairways, work areas, roadways, or other points of egress.
- Practice good housekeeping at all times.
- Training on equipment is required prior to unsupervised operation.

HOUSEKEEPING

- Proper housekeeping is the foundation for a safe work environment. It definitely helps prevent accidents and fires, as well as creating a professional appearance in the work area.
- Material will be piled or stored in a stable manner so that it will not be subject to falling.
- Combustible scrap, debris, and garbage shall be removed from the work area at frequent and regular intervals.
- Stairways, walkways, exit doors, in front of electrical panels, or access to fire-fighting equipment will be kept clear of materials, supplies, trash, and debris.

INDUSTRIAL HYGIENE AND OCCUPATIONAL HEALTH

- Toilet facilities shall be provided as required for the number of workers.
- An adequate supply of potable water shall be provided. The use of a common drinking cup is prohibited.
- Provisions will be made prior to commencement of the project for prompt medical attention in case of serious injury, to include emergency telephone numbers, transportation, and communications.

- When no medical facility is reasonably accessible (time and distance) to the worksite, a person who has a valid certificate of first aid training will be available at the worksite to render first aid.
- Employees must be protected against exposure to hazardous noise levels by controlling exposure or by use of proper personal protective equipment.
- Protection against exposure to harmful gases, fumes, dust, and similar airborne hazards must be furnished through proper ventilation or personal respiratory equipment.
- Any demolition work will be assessed for lead exposure (particularly if drywall, any painted surfaces, or abrasive blasting or grinding is involved) and asbestos exposure.

PERSONAL PROTECTIVE EQUIPMENT

- Personal protective equipment must be worn as required for each job in all operations where there is an exposure to hazardous conditions. Equipment requirements will be reviewed by the supervisor or foreman.
- Employees are expected to utilize proper judgment in their personal habits. When they report to work each morning they must be in fit condition to meet daily obligations.
- Goggles, face shields, helmets, and other comparable equipment are required to fit the eye and face protection needs of the employee for each job.
- Hard hats and steel-toed safety work boots or shoes must be worn by all employees at all times where required.
- Appropriate gloves, aprons, and boots are to be used when necessary for protection against acids and other chemicals that could injure employees' skin.
- Respiratory equipment in many cases is needed for protection against toxic and hazardous fumes or dusts. Supervisors must verify which equipment meets the need for breathing safety. Only Mine Safety and Health Administration (MSHA) and National Institute for Occupational Safety and Health (NIOSH)–approved equipment will be used.
- Some form or element of fall protection must be provided where employees are exposed to any fall hazard of six feet or greater (exceptions: scaffolds [ten feet] and ladders). Depending on the situation, this fall protection may be guardrails, safety nets, personal fall arrest systems (harness, lanyard, lifeline), hole covers, or any other appropriate protection.
- Flagmen will wear a red or orange warning garment while flagging; reflectorized garments will be worn at night.

Employers must review the Fall protection standard, 1926 Subpart M, for the various requirements for fall protection. Essentially, the standard requires that fall protection be addressed for any fall exposure over six feet.

ELECTRICAL

- Live electrical parts shall be guarded against accidental contact by cabinets, enclosure, location, or guarding. Cabinet covers will be replaced.
- Working and clear space around electric equipment and distribution boxes will be kept clear and accessible.
- Circuit breakers, switch boxes, and so forth, will be legibly marked to indicate their purpose.
- All 120-volt, single-phase 15- and 20-ampere receptacle outlets on construction sites, which are not a part of the permanent wiring of the building or structure and which are in use by employees, shall have approved ground-fault circuit interrupters (GFCI) for personnel protection. If the prime contractor has not provided this protection with GFCI receptacles at the temporary service drop, employees will ensure portable GFCI protection is provided. (Employers might wish to use an assured equipment grounding conductor program in lieu of this GFCI protection.) This requirement is in addition to any other electrical equipment grounding requirement or double-insulated protection.
- All extension cords will be three-wire (grounded) type and designed for hard or extra hard usage (Type S, ST, SO, STO, or SJ, SJO, SJT, SJTO). Ground prongs will not be removed. Cords and strain relief devices and clamps will be in good condition.
- All lamps for general illumination will have the bulbs protected against breakage. Temporary lights will not be suspended by their electrical cords unless cords and lights are designed for such suspension. Flexible cords used for temporary and portable lights will be designed for hard or extra hard usage.
- Employees will not work in such close (able to contact) proximity to any part of an electric power circuit unless the circuit is de-energized, grounded, or guarded by insulation.
- Equipment or circuits that are de-energized will be locked out and tagged out. The tags will plainly identify the equipment or circuits being worked on.

COMPRESSED GAS CYLINDERS

- All gas cylinders will have their contents clearly marked on the outside of each cylinder.
- Cylinders must be transported, stored, and secured in an upright position. They will never be left laying on the ground or floor, nor used as rollers or supports.
- Cylinder valves must be protected with caps and closed when not in use.
- All leaking or defective cylinders must be removed from service promptly, tagged as inoperable, and placed in an open space removed from the work area.
- Oxygen cylinders and fittings will be kept away from oil or grease.
- When cylinders are hoisted, they will be secured in a cradle, sling board, or pallet. Valve-protection caps will not be used for lifting cylinders from one vertical level to another.

LADDERS

- Ladders will be periodically inspected by a competent person to identify any unsafe conditions. Those ladders with structural defects will be removed from service and repaired or replaced.
- Straight ladders used on other than stable, level, and dry surfaces must be tied off, held, or secured for stability.
- Portable ladder side rails will extend at least three feet above the upper landing to which the ladder is used to gain access.
- The top or top step of a stepladder will not be used as a step.

AERIAL LIFTS

- Aerial lifts include cherry pickers, extensible boom platforms, aerial ladders, articulating boom platforms, vertical towers, and any combinations of the above.
- Only authorized and trained persons will operate aerial lifts.
- Lift controls will be tested each day before use.
- Safety harness will be worn when elevated in the aerial lift. Lanyards will be attached to the boom or basket. Employees will not belt off to adjacent poles, structures, or equipment while working from an aerial lift.
- Employees will always stand firmly on the floor of the basket and will not sit or climb on the edge of the basket. Planks, ladders, or other devices will not be used for work position or additional working height.
- Brakes will be set and outriggers will be used. The aerial lift truck will not be moved with the boom elevated and employees in the basket, unless the equipment is specifically designed for such use.

CRANES

- All cranes will be inspected by a competent person prior to each use and during use to make sure it is in safe operating condition. Also, a certification record of monthly inspections to include date, inspector signature, and crane identifier will be maintained.
- A thorough annual inspection of hoisting machinery will be made by a competent person, or by a government or private agency, and records will be maintained.
- Loads will never be swung over the heads of workers in the area.
- Employees will never ride hooks, concrete buckets, or other material loads being suspended or moved by cranes.
- Hand signals to crane operators will be those prescribed by the applicable American National Standards Institute (ANSI) standard to the type of crane in use.
- Tag lines must be used to control loads and keep workers away.
- Loads, booms, and rigging will be kept at least ten feet from energized electrical lines rated 50 KV or lower unless the lines are de-energized. For lines rated greater that 50 KV follow 1926.550(a)(15).

- Always operate cranes on firm, level surfaces, or use mats and pads, particularly for near capacity lifts.
- Accessible areas within the swing radius of the rear of the rotating superstructure of the crane, either permanently or temporarily mounted, will be barricaded in such a manner as to prevent employees from being struck or crushed by the crane.
- If suspended personnel platforms are to be lifted with a crane, reference 1926.550(g) for general and specific requirements.
- Rigging equipment (chains, slings, wire rope, hooks, other attachments, etc.) will be inspected prior to use on each shift to ensure it is safe. Defective rigging and equipment will be removed from service.
- Job or shop hooks or other makeshift fasteners using bolts, wire, and so forth, will not be used.
- Wire rope shall be taken out of service when one of the following conditions exist:
  - In running ropes, six random distributed broken wires in one lay or three broken wires in one strand or one lay;
  - Wear of one-third the original diameter of outside individual wires;
  - Kinking, crushing, bird caging, heat damage, or any other damage resulting in distortion of the rope structure; and
  - In standing ropes, more than two broken wires in one lay in sections beyond end connections, or more than one broken wire at an end connection.

WELDING AND BRAZING

- Combustible material will be cleared from the area around cutting or welding operations.
- Welding helmets and goggles will be worn for eye protection and to prevent flash burns. Eye protection to guard against slag while chipping, grinding, and dressing of welds will be worn.
- Only electrode holders specifically designed for arc welding will be used.
- All parts subject to electrical current will be fully insulated against the maximum voltage encountered to ground.
- A ground return cable shall have a safe current-carrying capacity equal to, or exceeding, the specified maximum output capacity of the arc welding unit that it services.
- Cables, leads, hoses, and connections will be placed so that there are no fire or tripping hazards.

TOOLS

- Take special precautions when using power tools. Defective tools will be removed from service.
- Electric power tools will be the grounded type or double insulated.

- Power tools will be turned off and motion stopped before setting tools down.
- Tools will be disconnected from power source before changing drills, blades, or bits, or attempting repair or adjustment. Never leave a running tool unattended.
- Power saws, table saws, and radial arm saws will have operational blade guards installed and used.
- Unsafe or defective hand tools will not be used. These include sprung jaws on wrenches, mushroomed head of chisels or punches, and cracked or broken handles of any tool.
- Portable abrasive grinders will have guards installed covering the upper and back portions of the abrasive wheel. Wheel speed ratings will never be less than the grinder RPM speed.
- Compressed air will not be used for cleaning purposes except when pressure is reduced to less than 30 psi by regulating or use of a safety nozzle, and then only with effective chip guarding and proper personal protective equipment.
- Abrasive blasting nozzles will have a valve that must be held open manually.
- Only trained employees will operate powder-actuated tools.
- Any employee furnished tools of any nature must meet all OSHA and ANSI requirements.

SAFETY RAILINGS AND FALL PROTECTION

- All open-sided floors and platforms six feet or more above adjacent floor or ground level will be guarded by a standard railing (top- and midrail, toeboard if required).
- A stairway or ladder will be provided at any point of access where there is a break in elevation of 19 inches or more.
- All stairways of four or more risers or greater than 30 inches high will be guarded by a handrail or stair rails.
- When a floor hole or opening (greater than two inches in its least dimension) is created during a work activity, through which a worker can fall or step into, or material can fall through, a cover or a safety guardrail must be installed immediately.
- Safety nets will be provided when workplaces are more than 25 feet above the ground, water, or other surfaces where the use of ladders, scaffolds, catch platforms, temporary floors, safety lines, or safety belts is impractical.
- Safety harnesses, lanyards, lines, and lifelines may be used in lieu of other fall protection systems to provide the required fall protection.
- Adjustment of lanyards must provide for not more than a six-foot fall, and all tie-off points must be at least waist high.

SCAFFOLDS

- Scaffolds will be erected, moved, dismantled, or altered only under the supervision of a competent person qualified in scaffold erection, moving, dismantling, or alteration.

- Standard guardrails (consisting of toprail and midrail) will be installed on all open sides and ends of scaffold platforms and work levels more than ten feet above the ground, floor, or lower level.
- Scaffolds four to ten feet in height with a minimum horizontal dimension in any direction less than 45 inches will have standard railings installed on all open sides and ends.
- Platforms at all working levels will be fully planked. Planking will be laid tight with no more than one inch of space between them, overlap at least 12 inches, and extend over end supports 6–12 inches.
- The front edge of all platforms will be no more than 14 inches from the face of the work, except plastering and lathing may be 18 inches.
- Mobile scaffolds will be erected no more than a maximum height of four times their minimum base dimension.
- Scaffolds will not be overloaded beyond their design loadings.
- Scaffold components should not be used as tie-off and anchor points for fall protection devices.
- Portable ladders, hook-on ladders, attachable ladders, integral prefabricated scaffold frames, walkways, or direct access from another scaffold or structure will be used for access when platforms are more than two feet above or below a point of access.
- Cross-braces will not be used as a mean of access to scaffolds.
- Scaffolds will not be erected, used, dismantled, altered, or moved such that they or any conductive material handled on them might come closer to exposed and energized power lines than the following:
    - Three feet from insulated lines of less than 300 volts; and
    - Ten feet plus for any other insulated or uninsulated lines.

EXCAVATIONS AND TRENCHES

- Any excavation or trench five feet or more in depth will be provided with cave-in protection through shoring, sloping, benching, or the use of hydraulic shoring, trench shields, or trench boxes. Trenches less than five feet in depth and showing potential of a cave-in will also be provided with cave-in protection. Specific requirements of each system are dependent upon the soil classification as determined by a competent person.
- A competent person will inspect each excavation and trench daily prior to the start of work, after every rainstorm or other hazard-increasing occurrence, and as needed throughout the shift.
- Means of egress will be provided in trenches four feet or more in depth so as to require no more than 25 feet of lateral travel for each employee in the trench.
- Spoil piles and other equipment will be kept at least two feet from the edge of the trench or excavation.

MOTOR VEHICLES AND MECHANIZED EQUIPMENT

- All vehicles and equipment will be checked at the beginning of each shift, and during use, to make sure it is in safe operating condition.

- All equipment left unattended at night adjacent to highways in normal use shall have lights or reflectors, or barricades with lights or reflectors, to identify the location of the equipment.
- When equipment is stopped or parked, parking brakes shall be set. Equipment on inclines shall have wheels chocked as well as having parking brakes set.
- Operators shall not use earthmoving or compaction equipment having an obstructed rear view unless the vehicle has an audible reverse signal alarm or is backed only when an observer says it is safe to do so.
- All vehicles shall have in operable condition the following:
  - ➢ Horn (bidirectional equipment);
  - ➢ Seats, firmly secured, for the number of persons carried (passengers must ride in seats);
  - ➢ Seat belts properly installed; and
  - ➢ Service, parking, and emergency brake system.

- All vehicles with cabs will be equipped with windshields with safety glass.
- All material handling equipment will be equipped with rollover protective structures.

MISCELLANEOUS

- All protruding reinforcing steel, onto and into which employees could fall, shall be guarded to eliminate the impalement hazard.
- Enclosed chutes will be used when material, trash, and debris are dropped more than 20 feet outside the exterior walls of a building. A substantial gate will be provided near the discharge end of the chute and guardrails at the chute openings into which workers drop material.
- Only trained employees will service large truck wheels. A cage or other restraining device plus an airline assembly consisting of a clip-on chuck, gauge, and length of hose will be used to inflate any large truck tires.
- Only trained employees will operate forklifts and other industrial trucks.

Safety and Health Programs

OSHA's Occupational Safety Regulations specify various individual programs that are applicable to our company. Highlights of these programs are provided below, and specific written programs or procedures are incorporated as appendixes into this document or are available separately.

COMPANY SAFETY RULES

These rules provide safety guidance for the company and employees to follow in the workplace. They cover various requirements in such areas as housekeeping, fire prevention, electrical, ladders, scaffolds, machine guarding, material handling, and so forth, that can be encountered in the workplace or on the job site.

HAZARD EVALUATION AND CONTROL PROGRAM

Employers are required to furnish to employees a workplace that is free from recognized hazards. An in-depth hazard evaluation or safety inspection conducted by OSHA's private consultants, insurance companies, or in-house personnel are means of identifying and eliminating workplace hazards. An ongoing periodic self-inspection program will help ensure that hazards are identified and eliminated or controlled.

HAZARD COMMUNICATION PROGRAM

If employees are exposed to or work with hazardous chemicals in the workplace, this program is required. Important elements of the program are as follows: a written program including a master listing of chemicals; material safety data sheets on each chemical; labeling; and training of employees on the program, on the chemicals exposed to, and on material safety data sheets.

PERSONAL PROTECTIVE EQUIPMENT HAZARD ASSESSMENT

Employers must assess their workplaces to determine if hazards are present, or are likely to be present, which necessitate the use or wear of personal protective equipment (eye and face, head, foot, or hand protection). This assessment must be documented through a written certification that identifies the workplace evaluated, the person certifying that the assessment has been completed, and the dates of the assessment; the document must also be identified as a certification of hazard assessment.

CONFINED SPACE ENTRY PROGRAM

If employees enter a confined space that contains or has the potential to contain an atmospheric or physical hazard, this program is required. Primary elements of the program are as follows: identification of applicable confined spaces, testing and monitoring, control or elimination of hazards, protective equipment, specific written entry authorization, attendants, training, and rescue.

RESPIRATORY PROTECTION PROGRAM

If employees are exposed to hazardous or toxic chemicals, paint or other gases, vapors, fumes, dusts, or mists above the permissible exposure limit, or respirators are worn by employees, this program is required. Program elements are as follows: written program for the selection, maintenance, care, and use of respirators; and fit testing, training, and employee physical evaluation for use.

OCCUPATIONAL NOISE EXPOSURE AND HEARING CONSERVATION PROGRAM

If employees are exposed to noise levels above the permissible noise exposures, protection against the effects of noise through engineering controls, administrative actions, or personal protective equipment, and an effective hearing conservation program, is required. Program elements would include a written program, identification and posting of hazardous noise areas, establishment of administrative actions for exposure control, noise monitoring, hearing evaluations and follow-on testing, personal protective equipment (hearing protection), and maintenance of medical records.

LOCKOUT OR TAGOUT PROGRAM

If employees service or maintain machines or equipment, and the unexpected energization or start-up of the equipment or release of stored energy could cause injury to the employee, this program is required. Such forms of hazardous energy include electrical, hydraulic, pneumatics, heat, or chemicals. Program elements include written energy control procedures delineating specific lockout or tagout action for each machine or piece of equipment, employee training, and periodic inspections.

EMERGENCY RESPONSE PLAN

If employees are engaged in emergency response to a hazardous substance or chemical release, an emergency response plan must be developed and implemented to handle anticipated emergencies. Program elements include a written response plan, identification and training of responding employees, medical surveillance and consultation, and post response operations.

CHEMICAL HYGIENE PLAN

A chemical hygiene program is required for those employees who work with chemicals in a laboratory. Program elements include a written plan, employee training, medical consultation and examinations, hazard identification, personal protective equipment, and recordkeeping.

EXPOSURE CONTROL PLAN

If employees are exposed to blood-borne pathogens or tuberculosis during the course of their work, this program is required. Program elements include a written plan, protective procedures and universal precautions, employee training, exposure incident treatment, and follow-up.

PROCESS SAFETY MANAGEMENT PROGRAM

If the company works in any way with certain highly hazardous chemicals in amounts above established threshold quantities, this program is required. Program elements include employee involvement, process identification, process hazard analysis, and establishment of operating procedures, employee training, a pre-start-up safety review, and incident investigations.

EMERGENCY ACTION PLAN

If required by a specific OSHA regulation (like Hazardous Waste Operations and Emergency Response [HAZWOPER]) for your company, this plan must be in writing (for companies with over ten employees) and cover those designated actions employers and employees must take to ensure employee safety from fire and other emergencies, such as a flood, tornado, and so on. Elements include response and evacuation procedures, alarming system, and training.

FIRE PREVENTION PLAN

If required by a specific OSHA regulation for your company, this plan must be in writing (for companies with over ten employees). Plan elements include major workplace fire hazards, housekeeping, and training.

ASBESTOS CONTROL PROGRAM

If employees are exposed to asbestos fibers in the workplace, then an initial monitoring for asbestos exposure must be made. If the monitoring results are above the permissible exposure limit (PEL), this program is required. Program elements include regulated areas, exposure monitoring, medical surveillance and records maintenance, engineering controls, personal protective equipment, and training.

LEAD EXPOSURE PROGRAM

If employees are exposed to lead in the workplace, then an initial monitoring for lead exposure must be made. If the monitoring results are above the permissible exposure limit (PEL), this program is required. Program elements include regulated areas, exposure monitoring, medical surveillance and records maintenance, engineering controls, personal protective equipment, and training.

Appendix A: Safety Meeting Minutes

Date: _____   Job name: _____

Topics: _____

_____

_____

_____

_____

_____

_____

_____

_____

_____

_____

_____

_____

Action items: _____

_____

_____

_____

_____

_____

_____

_____

_____

_____

_____

_____

_____

_____

_____

Meeting attended by:

_____   _____

_____   _____

_____   _____

_____   _____

_____   _____

Print name   Signature

Appendix B: Safety Program Policies and Procedures Acknowledgment

I have read and understand the attached company safety program and policies, and I agree to abide by them. I have also had the duties of the position that I have accepted explained to me, and I understand the requirements of the position. I understand that any violation of the above policies is reason for disciplinary action up to and including termination.

_____

Employee signature

_____

Date

Appendix C: Accident Investigation Report

An accident investigation form is not designed to find fault or place blame but is an analysis of the accident to determine causes that can be controlled or eliminated.

Company: _____

Address: _____

1. Name of injured: _____ S.S. #: _____
2. Sex: [ ] M [ ] F   Age: _____   Date of accident: _____
3. Time of accident: _____ a.m. _____ p.m.
4. Day of accident: _____
5. Employee's job title: _____
6. Length of experience on job: _____ (years) _____ (months)
7. Address of location where the accident occurred: _____
8. Nature of injury, injury type, and part of the body affected: _____
_____
_____

9. Describe the accident and how it occurred: _____
_____
_____

10. Cause of the accident: _____
_____
_____

11. Was personal protective equipment required? [ ] yes [ ] no
Was it provided? [ ] yes [ ] no
Was it being used? [ ] yes [ ] no   If "no," explain: _____
_____

Was it being used as you were trained by the supervisor or designated trainer?
[ ] yes [ ] no   If "no," explain: _____
_____
_____

12. Witnesses: _____
13. Was safety training provided to the injured? [ ] yes [ ] no
If "no," explain: _____
14. Interim corrective actions taken to prevent recurrence: _____
_____
_____

15. Permanent corrective action recommended to prevent recurrence: _____
_____
_____

16. Date of report: _____
Prepared by: _____
Supervisor (signature): _____
Date: _____

17. Status and follow-up action taken by safety coordinator: _____

_____

_____

Safety coordinator (signature): _____

Date: _____

Appendix D: Checklist for Evaluating Qualified Contractors

Name of contractor: _____

Contact person for contractor: _____

Title: _____

Address: _____

Telephone: _____

Please check items with the following: Yes or No

1. The contractor has obtained insurance coverage appropriate for the scope of work, prior to commencing the work (e.g., worker's compensation, general liability, etc.). (Attach certificates of insurance.)
    ___ Yes ___ No

2. The contractor has the necessary experience, references, and capability to properly perform the specific job at hand.
    ___ Yes ___ No

3. The contractor has a written safety program and agrees to conduct regular safety audits of its job sites by a competent person.
    ___ Yes ___ No

4. The contractor agrees to provide a site-specific safety plan including rigging, structural and RF safety procedures, and fall protection requirements for this specific job.
    ___ Yes ___ No

5. The contractor agrees there will be a competent and qualified person at the project site who will conduct daily safety audits.
    ___ Yes ___ No

6. The contractor agrees to maintain written records of the safety audits for a period of at least one year.
    ___ Yes ___ No

7. The contractor requires preemployment physical agility or physical fitness tests to determine ability to perform job tasks.
    ___ Yes ___ No

8. The contractor conducts drug screening of employees for unlawful use of controlled substances.
    ___ Yes ___ No

9. The contractor provides an orientation and awareness program for new hires prior to performance of any work.
    ___ Yes ___ No

10. The contractor ensures that the tower climbers have been properly trained and understand OSHA regulations in the areas of fall protection and rescue.
    ___ Yes ___ No

11. The contractor agrees to conduct a hazard assessment to determine the requirements for personal protective equipment, including fall protection.
  ___ Yes ___ No
12. The contractor maintains written documentation of all training as required.
  ___ Yes ___ No
13. If the contractor is required to maintain OSHA 300 logs, he or she has submitted those documents for the past two years. If not required to keep OSHA 300 logs, the company has provided the number of employees and a report on accidents sustained, including the nature, type, and number of accidents for the past two years.
  ___ Yes ___ No
14. The contractor agrees to notify the company in writing if subcontractors are to be used prior to the use of such subcontractors.
  ___ Yes ___ No
15. The contractor agrees that any subcontractors hired will be required to meet the same contractor requirements outlined in this document.
  ___ Yes ___ No
16. The contractor agrees to maintain good housekeeping on the job site.
  ___ Yes ___ No

Individual completing questionnaire: _____

Title: _____

Date: _____

This document will be kept on file in the safety manager's office.

Index